Lecture Notes in Statistics

W0111314

Vol. 1: R. A. Fisher: An Appreciation. Edited by S. E. Fienberg and D. V. Hinkley. xi, 208 pages, 1980.

Vol. 2: Mathematical Statistics and Probability Theory. Proceedings 1978. Edited by W. Klonecki, A. Kozek, and J. Rosiński. xxiv, 373 pages, 1980.

Vol. 3: B. D. Spencer, Benefit-Cost Analysis of Data Used to Allocate Funds. viii, 296 pages, 1980.

Vol. 4: E. A. van Doorn, Stochastic Monotonicity and Queueing Applications of Birth-Death Processes. vi, 118 pages, 1981.

Vol. 5: T. Rolski, Stationary Random Processes Associated with Point Processes. vi, 139 pages, 1981.

Vol. 6: S. S. Gupta and D.-Y. Huang, Multiple Statistical Decision Theory: Recent Developments. viii, 104 pages, 1981.

Vol. 7: M. Akahira and K. Takeuchi, Asymptotic Efficiency of Statistical Estimators. viii, 242 pages, 1981.

Vol. 8: The First Pannonian Symposium on Mathematical Statistics. Edited by P. Révész, L. Schmetterer, and V. M. Zolotarev. vi, 308 pages, 1981.

Vol. 9: B. Jørgensen, Statistical Properties of the Generalized Inverse Gaussian Distribution. vi, 188 pages, 1981.

Vol. 10: A. A. McIntosh, Fitting Linear Models: An Application of Conjugate Gradient Algorithms. vi, 200 pages, 1982.

Vol. 11: D. F. Nicholls and B. G. Quinn, Random Coefficient Autoregressive Models: An Introduction. v, 154 pages, 1982.

Vol. 12: M. Jacobsen, Statistical Analysis of Counting Processes. vii, 226 pages, 1982.

Vol. 13: J. Pfanzagl (with the assistance of W. Wefelmeyer), Contributions to a General Asymptotic Statistical Theory. vii, 315 pages, 1982.

Vol. 14: GLIM 82: Proceedings of the International Conference on Generalised Linear Models. Edited by R. Gilchrist. v, 188 pages, 1982.

Vol. 15: K. R. W. Brewer and M. Hanif, Sampling with Unequal Probabilities. ix, 164 pages, 1983.

Vol. 16: Specifying Statistical Models: From Parametric to Non-Parametric, Using Bayesian or Non-Bayesian Approaches. Edited by J. P. Florens, M. Mouchart, J. P. Raoult, L. Simar, and A. F. M. Smith. xi, 204 pages, 1983.

Vol. 17: I. V. Basawa and D. J. Scott, Asymptotic Optimal Inference for Non-Ergodic Models. ix, 170 pages, 1983.

Continued

Lecture Notes in Statistics

Vol. 18: W. Britton, Conjugate Duality and the Exponential Fourier Spectrum. v, 226 pages, 1983.

Vol. 19: L. Fernholz, von Mises Calculus for Statistical Functionals. viii, 124 pages, 1983.

Vol. 20: Mathematical Learning Models—Theory and Algororithms: Proceedings of a Conference. Edited by U. Herkenrath, D. Kalin, W. Vogel. xiv, 226 pages, 1983.

Vol. 21: H. Tong, Threshold Models in Non-linear Time Series Analysis. x, 323 pages, 1983.

Lecture Notes in Statistics

Edited by D. Brillinger, S. Fienberg, J. Gani,
J. Hartigan, and K. Krickeberg

20

Mathematical Learning Models—Theory and Algorithms

Proceedings of a Conference

Edited by
Ulrich Herkenrath, Dieter Kalin, and
Walter Vogel

Springer Science+Business Media, LLC

Ulrich Herkenrath
Institut für Angewandte Mathematik
der Universität Bonn
Wegelerstrasse 6
5300 Bonn
Federal Republic of Germany

Dieter Kalin
Universität Ulm,
Abt. Mathematik VII
Oberer Eselsberg
7900 Ulm
Federal Republic of Germany

Walter Vogel
Institut für Angewandte Mathematik
der Universität Bonn
Wegelerstrasse 6
5300 Bonn
Federal Republic of Germany

AMS Subject Classifications: 62-06, 62L20, 62L99

Library of Congress Cataloguing in Publication Data
Mathematical learning models—theory and algorithms.
(Lecture notes in statistics; 20)
"Organized by the Institute of Applied Mathematics of the University of Bonn . . . [and held] in the Physikzentrum in Bad Honnef . . . from May 3–May 7, 1982"—Pref.
1. Learning models (Stochastic processes)—Congresses.
I. Herkenrath, Ulrich. II. Kalin, Dieter. III. Vogel, Walter, 1923 June 22– . IV. Universität Bonn. Institut für Angewandte Mathematik. V. Series: Lecture notes in statistics (Springer-Verlag); v. 20.
QA274.6.M37 1983 519.2 83-17011

With 4 Illustrations

9 8 7 6 5 4 3 2 1

ISBN 978-0-387-90913-4 ISBN 978-1-4612-5612-0 (eBook)
DOI 10.1007/978-1-4612-5612-0

Preface

This volume contains most of the contributions presented at the conference "Mathematical Learning Models - Theory and Algorithms". The conference was organized by the Institute of Applied Mathematics of the University of Bonn under the auspices of the Sonderforschungsbereich 72. It took place in the Physikzentrum in Bad Honnef near to Bonn from May 3 - May 7, 1982.

The idea of the organizers was to bring together experts who work on very related problems, but partially by using different approaches. The main subjects of the program were:

- mathematical learning models,
- bandit problems,
- stochastic approximation procedures,
- sequential decision processes with unknown law of nature.

We felt that in a sense "learning" was a common concept for all these branches.

In the contributions the state of the art in the above topics was presented from different points of view with special regard to recent advances. The exchange of results and opinions was continued in many fruitful and vivid discussions. The atmosphere of the conference center offered a suitable and pleasant framework for the scientific program.

We express our gratitude to all contributors for making the conference successful. Simultaneously we hope that further work on the above mentioned field has been stimulated.

We gratefully acknowledge the financial support of the Sonderforschungsbereich 72 of the Deutsche Forschungsgemeinschaft and the hospitality of the Physikzentrum in Bad Honnef. Last not least we would like to thank I. Kreuder, M. Thiedemann and A. Thiedemann for their attentive assistance in organizing the conference.

U. Herkenrath D. Kalin W. Vogel

Introduction

The aim of the organizers of this conference was, to bring together scientists who work on seemingly different branches of modern stochastics, but on problems with a common basis. These different branches may be sketched by the key words: mathematical learning models, bandit problems, stochastic approximation procedures, sequential decision processes with unknown law of nature. The common basis of the topics dealt at the conference, may be described as follows:

Let A be a set of alternatives (= arms of the bandit, parameters of a stochastic approximation model), where each alternative $a \in A$ is represented by a chance experiment. The chance experiment is described by a real-valued random variable $Y(a)$ with distribution function F_a and expected value $R(a)$. The outcomes of $Y(a)$ are interpreted as gains and the F_a's and expected gains $R(a)$ are unknown to a learning subject (or statistician) with the exception of perhaps some structural properties. The learning subject is confronted with these data in the sense, that in each period of time $n = 1,2,3,\ldots$ it has to choose one of the alternatives $a \in A$, i.e. to observe one of the $Y(a)$'s. Of course it has the aim to select the alternatives in such a way, that it finds out in course of time the alternative Θ, where R as a function on A attains a certain value (e.g. the maximal one) and simultaneously to obtain a reasonable gain by observing different $Y(a)$'s. This causes a conflict between the two goals of improving information on the unknown value Θ and of maximizing a certain objective concerning the gains, which are earned from the observed alternatives. Some authors denote this as the conflict between estimation and control.

This general problem may be formalized in various models and studied under different aspects and criteria. Perhaps the concept of "learning" is common to all approaches to the above problem, wherefore we integrated this word into the title of the conference and of this volume. Of course there is no unambiguous use of this word in the literature, but we feel this term justified in the case, that the only way to improve information is to participate in a game, where one wants to behave in an "optimal" fashion with respect to some objective. The variety of concrete models fitting into the above general framework is reflected in the spectrum of the contributions to the conference and to this volume.

There is one group of contributions, which deal with models and techniques, in which the term "learning" appears directly. Iosifescu and Pruscha study general models for the analysis of learning processes, the first one limit theorems for time-discrete models and the latter one the construction of time-continuous models. Learning algorithms within the so-called reinforcement schemes are analyzed by Lakshmivarahan, Meybodi. In El-Fattah's paper a learning automaton is presented to tackle a semi-Markov decision process.

A second group of papers refers to the well-known bandit problem. Bather studies the minimax risk for the two-armed bandit and gives a lower asymptotic bound for this term. Berry analyzes a k-armed bandit with random discounting, for which the expected sum of the weighted observations is maximized. Gittins and Glazebrook present new results on the dynamic allocation index (nowadays already known as Gittins index) and in particular its application to Bayesian bandits. A Bayesian framework is also chosen by Jones, Kandeel to perform numerical studies for optimal policies. Siegmund pursues in his contribution mainly the applicability of the two-armed bandit for the design of sequential clinical trials. Finally Witten demonstrates in his paper, how a two-armed bandit controller may be employed as a basic instrument to solve more complex systemtheoretic problems.

A further group of articles is devoted to problems of stochastic approximation procedures. Dupač, Fiala and Monnez study convergence properties of stochastic approximation algorithms under convex constraints. Fabian demonstrates an optimality property of an adaptive Robbins-Monro procedure and Pflug derives some convergence properties for a Robbins-Monro procedure. Ruppert turns toward a modification of the classical models, which has attracted considerable attention in recent time, the case of dependent disturbances. Lai applies stochastic approximation techniques to stochastic regression models. Finally Ljung describes in his paper how methods of stochastic approximation fit into system theory.

There remain some papers which deal with special problems arising in the control of stochastic processes with unknown parameters in the law of motion. This is for the case of the Wiener process the paper of Helmes, for stopping problems Irle's article and for time-discrete Markovian decision processes the contribution of Kolonko. Concerning unknown distribution parameters occuring in a Markov-chain the paper of Schmitz, Süselbeck discusses some aspects of the sequential probability ratio test. Mandl deals in his paper with an optimization problem for matrices, which has an application to decision models.

We hope that this collection of papers stimulates further research on each of the above topics and moreover contributes to a mutual fructification of the different branches.

CONTENTS:

Papers presented at the conference but not received for the Proceedings:

K. MARTI
Controlled Random Search Procedures in Optimization: Optimal Controls and Rates of Convergence

G. N. SARIDIS
Hierarchically Intelligent Control Systems

M. SCHÄL
Recursive Estimation in Dynamic Programming

H.-J. SEBASTIAN
On the Theory of Optimal Control of Time-Discrete Processes with Incomplete Information with Regard to Learning Abilities

M. SOBEL
Sequential Allocation Rules

R. THEODORESCU
On a Class of Linear Models

Y. Z. TSYPKIN
Optimality Under Uncertainty

H. WALK
On Recursive Estimation of the Mode

H. WOLFF
Learning Time-Dependent Densities

LIST OF PARTICIPANTS

Abdel-Fattah,Y.M.
Laboratory for Electronics
and System Control
Faculty of Science
University Mohamed
Rabat - Marocco

Bather,J.A.
Mathematics Division
University of Sussex
Falmer,Brighton BN1 9QH
Great Britain

Berry,D.A.
Dept.of Theoretical Statistics
University of Minnesota
27o Vincent Hall
2o6 Church Str. S.E.
Minneapolis,Minnesota 55455
USA

Bert,M.C.
U.E.R. de Mathématiques
Université de Paris V
1o Résidence du Chateau
91.19o Gif-Sur-Yvette
France

Bodewig,H.H.
Institut für Angew.Mathematik
Universität Bonn
Wegelerstr.6
D - 53oo Bonn

Bosch,K.
Institut für Angew.Mathematik
Universität Hohenheim
Postfach 7oo562/11o
D - 7ooo Stuttgart 7o

Daniel,K.
Institut für Mathem.Statistik
Universität Bern
Sidlerstr.5
CH - 3o12 Bern

Dawen van,R.
Institut für Angew.Mathematik
Universität Bonn
Wegelerstr.6
D - 53oo Bonn

Dupač,V.
Dept.of Probability&Statistics
Charles University
Sokolovska 83
186oo Prague 8
CSSR

Ersü,E.
FG Regelsystemtheorie
TH Darmstadt
Schloßgraben 1
D - 61oo Darmstadt

Fabian,V.
Dept.of Statistics&Probability
Michigan State University
East Lansing
Michigan 48824
USA

Gittins,J.C.
Mathematical Institute
Oxford University
25-29 St.Giles
Oxford OX1 3LB
Great Britain

Glazebrook,K.
Dept.of Statistics
University of Newcastle-Upon-Tyne
Newcastle-Upon-Tyne NE1 7RU
Great Britain

Goldmann,W.
Institut für Angew.Mathematik
Universität Bonn
Wegelerstr.6
D - 53oo Bonn

Hartung,J.
Abteilung Statistik
Universität Dortmund
Postfach 5oo5oo
D - 46oo Dortmund

Helmes,K.
Institut für Angew.Mathematik
Universität Bonn
Wegelerstr.6
D - 53oo Bonn

Herkenrath,U.
Institut für Angew.Mathematik
Universität Bonn
Wegelerstr.6
D - 53oo Bonn

Hinderer,K.
Institut für Mathem.Statistik
Universität Karlsruhe
Englerstr.2
D - 75oo Karlsruhe 1

Iosifescu,M.
Centre of Mathem.Statistics
str.Stirbei Voda 174
771o4 Bucharest
Romania

Irle,A.
Institut für Mathem.Statistik
Universität Münster
Einsteinstr.62
D - 44oo Münster

Jones,P.W.
Dept.of Mathematics
University of Keele
Staffordshire ST5 5BG
Great Britain

Kalin,D.
Abteilung Mathematik VII
Universität Ulm
Oberer Eselsberg
D - 79oo Ulm

Klinger,H.
Institut für Statistik
Universität Düsseldorf
Universitätsstr.1
D - 4ooo Düsseldorf

Kolonko,M.
Institut für Mathem.Statistik
Universität Karlsruhe
Englerstr.2
D - 75oo Karlsruhe

Lai,T.L.
Columbia University
618 Mathematical Building
New York 1oo27
USA

Lakshmivarahan,S.
School of Electrical Engineering
and Computer Science
University of Oklahoma
2o2 West Boyd
Norman,Oklahoma 73o19

Lehn,J.
Fachbereich Mathematik
TH Darmstadt
Schloßgartenstr.7
D - 61oo Darmstadt

Lerche,R.
Institut für Angew.Mathematik
Universität Heidelberg
Im Neuenheimer Feld 294
D - 69oo Heidelberg

Ljung,L.
Dept.of Electrical Engineering
Linköping University
S - 581 81 Linköping

Mammitzsch,V.
Fachbereich Mathematik
Universität Marburg
Lahnberge
D - 355o Marburg/Lahn

Mandl,P.
Dept.of Probability&Statistics
Charles University
Sokolovska 83
186oo Prague 8
CSSR

Mann,E.
Institut für Angew.Mathematik
Universität Bonn
Wegelerstr.6
D - 53oo Bonn

Marti,K.
Fachbereich Luft- und
Raumfahrttechnik
Hochschule der Bundeswehr München
Werner-Heisenberg-Weg 39
D - 8o14 Neubiberg

Meyer zu Hörste,U.
Institut für Angew.Mathematik
Universität Bonn
Wegelerstr.6
D - 53oo Bonn

Monnez,J.M.
I.U.T. Informatique
Université de Nancy II
Boulevard Charlemagne
F - 54ooo Nancy

Pflug,G.
Institut für Mathematik
Universität Gießen
Arndtstr.2
D - 63oo Gießen

Postelnicu,T.
Academy Rep. Soc. Romania
Calea Victoriei Nr.125
771o4 Bucharest
Romania

Pruscha,H.
Institut für Mathematik
Universität München
Theresienstr.39
D - 8ooo München

Rauhut,B.
Institut für Statistik
und Wirtschaftsmathematik
RWTH Aachen
Wüllnerstr.3
D - 51oo Aachen

Reimnitz,P.
Institut für Med.Statistik
Universität Bonn
Venusberg
D - 53oo Bonn

Rieder,U.
Abteilung Mathematik VII
Universität Ulm
Oberer Eselsberg
D - 79oo Ulm

Rudolph,A.
Institut für Angew.Mathematik
Universität Bonn
Wegelerstr.6
D - 53oo Bonn

Ruppert,D.
Dept.of Statistics
University of North Carolina
Phillips Hall o39A
Chapel Hill, NC 27514
USA

Sänger,G.
Institut für Angew.Mathematik
Universität Bonn
Wegelerstr.6
D - 53oo Bonn

Saridis,G.N.
Dept.of ESCE
Rensselaer Polytechnic Institute
Troy, NY 12181
USA

Schäl,M.
Institut für Angew.Mathematik
Universität Bonn
Wegelerstr.6
D - 53oo Bonn

Schellhaas,H.
Fachbereich Mathematik
TH Darmstadt
Schloßgartenstr.7
D - 61oo Darmstadt

Schmitz,N.
Institut für Mathem.Statistik
Universität Münster
Einsteinstr.62
D - 44oo Münster

Sebastian,H.J.
Sektion Mathematik
TH Leipzig
Karl-Liebknecht-Str.132
DDR - 7o3o Leipzig

Siegmund,D.
Dept.of Statistics
Stanford University
Stanford CA 943o5
USA

Sobel,M.
Dept.of Mathematics
UC Santa Barbara
Santa Barbara CA 931o6
USA

Theodorescu,R.
Département de Mathématiques
Université Laval
Québec G1K 7P4
Canada

Tsypkin,Y.Z
Institute of Control Sciences
USSR Academy of Sciences
65 Profsoyuznaya
Moscow 117342
USSR

Vogel,W.
Institut für Angew.Mathematik
Universität Bonn
Wegelerstr.6
D - 53oo Bonn

Waldmann,K.H.
Fachbereich Mathematik
TH Darmstadt
Schloßgartenstr.7
D - 61oo Darmstadt

Walk,H.
Institut für Mathematik
Universität Gießen
Arndtstr.2
D - 63oo Gießen

Witten,I.H.
Dept.of Computer Science
University of Calgary
Calgary T2N 1N4
Canada

Wolff,H.
Institut für Angew.Mathematik
TU Braunschweig
Pockelstr.14
D - 33oo Braunschweig

THE MINIMAX RISK FOR THE TWO-ARMED BANDIT PROBLEM

J.A. Bather, University of Sussex,
Falmer, Brighton, U.K.

Let r be the number of successes scored in t independent trials, each of which must be allocated to one of two experiments, where the probabilities p_1, p_2 of success in these experiments are both unknown. For any allocation rule Δ, the expected number of successes lost as a result of ignorance is

$$L_\Delta(p_1,p_2,t) = t\max(p_1,p_2) - E_\Delta(r).$$

A minimax rule is one that attains

$$M(t) = \inf_{\Delta}\ \sup_{p_1,p_2} L_\Delta(p_1,p_2,t).$$

Such rules exist for any given integer t, but they are very difficult to find unless t is small. However, Vogel (1960) obtained asymptotic bounds on the minimax risk:

$$0.187 \le \liminf_{t\to\infty} M(t)/t^{\frac{1}{2}} \le \limsup_{t\to\infty} M(t)/t^{\frac{1}{2}} \le 0.376.$$

Since then, some progress has been made in improving these bounds and the main result described here is that

$$\liminf_{t\to\infty} M(t)/t^{\frac{1}{2}} \ge 0.306.$$

We now know roughly what is possible, at least when t is large, but it is still not very clear how to achieve this.

1. Introduction.

Two-armed and multi-armed bandit problems have been studied quite intensively during the last thirty years. The association with gambling machines is slightly misleading, since the most important application is in sequential medical trials. Imagine a sequence of t patients and two or more possible treatments for any one of them. The problem is to determine a rule for allocating a treatment to each patient depending, at each stage, only on previously observed successes and failures. In the first mathematical investigation, Robbins (1952) considered two alternative treatments and we shall concentrate on this case.

Many different allocation rules have been proposed, but most of them are inefficient with respect to the minimax criterion:

$$M_\Delta(t) = \sup_{p_1,p_2} L_\Delta(p_1,p_2,t).$$

In particular, it is worth noting that a simple test which leads to a final decision that $p_1>p_2$ or $p_1<p_2$ after a fixed sample of, say, 50 observations on each treatment must have a maximum risk $M_\Delta(t)$ of order t. If the number of patients is known in advance, the sample sizes $s_1 = s_2$ can be chosen to depend on t and a straightforward calculation shows that the best possible result for rules of this type is that $M_\Delta(t) \simeq 0.505t^{\frac{1}{2}}$ when t is large.

Vogel's asymptotic upper bound on $M(t)$ is based on using a sequential probability ratio test for two hypotheses of the form

$$H_1 : p_1 = \tfrac{1}{2}(1+\delta),\ p_2 = \tfrac{1}{2}(1-\delta); \qquad H_2 : p_1 = \tfrac{1}{2}(1-\delta),\ p_2 = \tfrac{1}{2}(1+\delta),$$

where $\delta = |p_1-p_2|$ is fixed. He restricted attention to a simple allocation rule with pairs of observations, one on each treatment, during the period of testing. The difference r_1-r_2 between the numbers of successes obtained with p_1 and p_2 is recorded after each pair and, finally, treatment 1 or 2 is preferred for all the remaining patients as soon as $r_1-r_2 = \pm D$, where the critical level $D = D(t)$. It is not so easy to evaluate this kind of policy, but Vogel established that one can choose D and hence find a procedure $\Delta = \Delta(t)$ such that $M_\Delta(t)/t^{\frac{1}{2}} \to 0.376$ as $t \to \infty$.

This result shows that a substantial reduction in the maximum risk can be obtained by using a sequential stopping rule. Further reductions can be achieved by using sequential allocations, but the policies are more complicated and they are very difficult to evaluate without extensive simulations. Two possible approaches will be mentioned in order to illustrate the possibilities: we now know enough to devise reasonably effective policies but, from a practical point of view, it is important to develop and promote allocation rules which are both simple and effective.

The idea of using a separate index of performance for each of the arms or treatments in a multi-armed bandit process can be helpful. This principle has been justified , under certain assumptions, for a wide range of allocation problems: see Gittins (1979) and also his contribution to the present conference. In particular, suppose there are k treatments with unknown success probabilities $p_1, p_2, \ldots, p_k$, represented by independent prior distributions $\pi_1, \pi_2, \ldots, \pi_k$. Let r_i be the number of successes observed in s_i trials with treatment i, after a total of $s = \Sigma s_i$ trials. Consider a Bayesian formulation of the problem with a discount factor β, $0<\beta<1$, and an infinite sequence of patients. Under these conditions, the optimal policy is determined by Gittins' dynamic allocation indices G_i, $i = 1,2,\ldots,k$, at each stage. Here, $G_i = G_i(r_i, s_i, \pi_i, \beta)$ depends only on the information accumulated about treatment i at any particular time s. The Bayes risk is minimised by applying the following allocation rule throughout: use treatment j for the next patient if and only if j is the first integer such that $G_j = \max\{G_1, G_2, \ldots, G_k\}$. Roughly speaking, this means that discounted Bayesian versions of the multi-armed bandit problem can be solved by computing the solutions of separate one-armed bandit problems to determine the functions G_i. This is a considerable simplification, even when $k=2$, but it requires extensive computations and one difficulty is that the indices are highly sensitive to the choice of discount factor, especially when β is near 1.

My own interest in bandit problems began with the notion that simple and effective policies might be developed by using randomised allocation indices: see Bather (1980) and (1981). It is unrealistic to assume, for example, in the finite horizon version of the two-armed bandit problem, that the total t of all the patients involved is specified in advance. This led to a study of stationary decision procedures, not depending on t, and also to an invariance principle based on approximating the long term behaviour of allocation rules for Bernoulli sequences by suitable diffusion models: see Section 4 of the (1981) paper. For our present purpose, it is enough to remark that a useful policy emerged after many simulations and comparisons with other allocation rules. The policy is based on indices defined in the following way:

$$Q_i = \frac{r_i}{s_i} + \frac{(4 + s_i^{\frac{1}{2}})}{15 s_i} \{2 + \xi_i(s)\} ,$$

where the random variables $\xi_i(s)$, $i = 1,2,\ldots,k$, are independent of one another and all previous events, with a common probability density $e^{-\xi}$, $\xi>0$. As before, the treatments must be allocated so that the next patient always received the one having the largest index. This policy is stationary and its maximum risk $M_Q(t)$ is of order $t^{\frac{1}{2}}$ for large t. In fact, the simulations confirmed that $M_Q(t)/t^{\frac{1}{2}} \approx 0.36$ for $t \geq 50$. Thus, a single policy can achieve a slightly better result, for all large t, than any of the family of allocation rules derived from the sequential probability ratio test.

Let us return to the assumption that the horizon t is given. Another possible approach is to compute Bayes solutions for particular choices of prior distribution and then evaluate the maximum of the corresponding risk functions by using simulations. For example, suppose that p_1 and p_2 have independent, uniform prior distributions and let $t = 100$. The Bayes procedure can be determined in this case by standard dynamic programming techniques but, so far as I am aware, the maximum risk has not been evaluated. However, a similar investigation has been carried out recently by Robinson (1982), using an approximation to the Gittins index rule for the above prior distributions and a discount factor β chosen so that $(1-\beta)^{-1} = t$. He found that the allocation rule $G = G(t)$ has a maximum risk $M_G(t) \approx 3.2$ when $t = 100$. The invariance principle mentioned earlier enables one to transform allocation rules from one time interval to another, by using normal approximations. Hence, the special case indicates that $M_G(t)/t^{\frac{1}{2}} \approx 0.32$ when t is large.

In principle, one can find lower bounds on the minimax risk $M(t)$ by noting that the Bayes risk $B_\pi(t)$ obtained by minimising the expected loss with respect to any prior distribution π on p_1, p_2 always has the property that $B_\pi(t) \leq M(t)$. Vogel (1960) used this approach and the same idea, with a different choice of prior distribution, was used by Fabius and van Zwet (1970) to strengthen his asymptotic lower bound. They established that

$$\liminf_{t \to \infty} M(t)/t^{\frac{1}{2}} \geq 0.265.$$

This and some other results from their paper will be described briefly in the next section.

We can get closer to the true behaviour of $M(t)$ when t is large, by introducting a modification of the two-armed bandit problem. This leads to an artificial situation in which one can separate the aim of sampling for information about p_1 and p_2 from the need to give patients the benefit of the more promising of the two treatments at each stage. Section 3 explains this modified sequential decision problem and it contains an outline of the argument leading to the result that

$$\liminf_{t \to \infty} M(t)/t^{\frac{1}{2}} \geq 0.306.$$

It must be emphasised that the decision procedures involved in this argument offer very little guidance towards minimax allocation rules for the original problem.

At the meeting in Bad Honnef, I stated a figure of 0.283 in the last inequality and conjectured a further improvement, partly as a result of contemplations during the journey. Since then, I have been able to confirm the improvement, at least, up to a rather complicated diffusion approximation. A detailed proof of the result will be published later.

2. Lower bounds on the minimax risk.

The main purpose of the paper by Fabius and van Zwet (1970) was to establish the existence of a minimax strategy for the two-armed bandit problem, for each integer $t \geq 1$. They derived certain monotonicity properties which apply to any admissible strategy and used these to determine the minimax rule for $t \leq 4$, remarking that the algebra involved becomes progressively more complicated and seems prohibitive even when $t=5$. The results for small t are instructive and they suggest, together with the asymptotic bounds already mentioned, that the ratio $M(t)/t^{\frac{1}{2}}$ converges rapidly to a constant:

t	=	1	2	3	4
$M(t)/t^{\frac{1}{2}}$	=	.5000	.3536	.3248	.3087 .

The paper also shows, without formal proof, how to improve Vogel's asymptotic lower bound and, since the method provides a clue to further improvements, it is worth explaining. Consider the prior distribution defined by assigning equal probabilities to the simple hypotheses

$$H_1 : p_1 = \tfrac{1}{2}(1+\delta),\ p_2 = \tfrac{1}{2}(1-\delta); \quad H_2 : p_1 = \tfrac{1}{2}(1-\delta),\ p_2 = \tfrac{1}{2}(1+\delta),$$

where δ is given, $0<\delta<1$. Since $p_1 + p_2 = 1$ under either hypothesis it is clear that a success observed in any trial with p_1 is equivalent to a failure with p_2 and vice-versa. In fact, the information about p_1 and p_2 derived from any sequence of trials does not depend on the sampling rule. It is easy to see that the Bayes procedure is determined by the following rule: use p_1 in the next trial if $2r_1 - s_1 > 2r_2 - s_2$ and p_2 if the inequality is reversed; choose p_1 or p_2 by tossing a coin in case of equality. The difference $z(s) = (2r_1-s_1) - (2r_2-s_2)$ has a distribution which is approximately normal when $s = s_1 + s_2$ is large and, for example, if H_1 is true, the distribution has mean δs and variance $(1-\delta^2)s$. It is a straightforward exercise to approximate the Bayes risk $B_\pi(t,\delta)$ and use the fact that $M(t) \geq \sup B_\pi(t,\delta)$, taking the supremum with respect to δ. This leads to the result that

$$\liminf_{t \to \infty} M(t)/t^{\frac{1}{2}} \geq 0.265.$$

One can see roughly where this argument gives too much away. The fact that the prior distribution is concentrated on points where $p_1 + p_2 = 1$ means that, in effect, the decision maker may assume, in calculating the probability of choosing the inferior treatment at any time s, that he has previously observed samples of equal size $s_1 = s_2 = \frac{1}{2}s$ on each treatment. This is unrealistic and it is worth considering a modification of the problem which retains the feature of separating the sampling rule from the allocation rule, without permitting too many observations on the inferior treatment.

3. A modified decision problem.

This final section contains an outline of the proof that the minimax risk $M(t)$ for the two-armed bandit problem with t trials has the property that

$$\text{(1)} \qquad \liminf_{t \to \infty} M(t)/t^{\frac{1}{2}} \geq .306.$$

We start with a modified version of the decision problem, involving four independent sequences of independent Bernoulli trials instead of two. These will be denoted by I_1, J_1 and I_2, J_2, where the underlying probability of success for each single trial is p_1 in the first two sequences and p_2 in the others. The idea is that the decision maker may sample observations from either I_1 or I_2 and then, by using this information sequentially, he must choose a succession of treatments from J_1 or J_2. The results of trials selected from J_1 and J_2 will not be revealed to the decision maker until all the trials are completed, say, at time $t+1$. A decision rule $\Delta = (\Delta(I), \Delta(J))$ consists of two parts: the sampling rule $\Delta(I)$, which indicates whether the next observation is taken from I_1 or I_2, and the allocation rule $\Delta(J)$. Thus, after s observations, including s_1 from I_1 and s_2 from I_2, treatment 1 or 2 is allocated to the $(s+1)$th patient by using this information according to $\Delta(J)$.

Of course, this model is highly unrealistic, but we shall impose a restriction on the decision rules which brings it closer to the original problem. For any rule Δ, consider the random variables $m(s)$ and $n(s)$, $s = 1,2,\ldots,t$, where $m(s) = \min(s_1,s_2)$ and $n(s)$ is the number of allocations of the inferior treatment amongst the first s selections from J_1,J_2. The expected loss, equivalent to $L_\Delta(p_1,p_2,t)$, can be evaluated as $|p_1-p_2|\ E_\Delta(n(t)|p_1,p_2)$ and for example, if $p_1>p_2$, $n(t)$ is the number of trials selected from J_2. We now define a class $C_0 = C_0(t)$ of admissible decision procedures by saying that $\Delta \in C_0$ if

(2) $$E_\Delta(m(s)|p_1,p_2) \le E_\Delta(n(s)|p_1,p_2)$$

for each $s \le t$. It will be convenient to consider pairs of probabilities with $|p_1 - p_2| = \delta$, where δ is fixed, so we define the corresponding minimax risk as

(3) $$M_0(t,\delta) = \inf_{\Delta \in C_0} \delta \sup_{|p_1 - p_2| = \delta} E_\Delta(n(t))|p_1,p_2).$$

Suppose that Δ is the minimax rule for the two-armed bandit: this can be applied to the modified problem simply by demanding that $\Delta(I) = \Delta(J)$, in the sense that any observation from I_1 must lead to a simultaneous allocation from J_1 and similarly for I_2 and J_2. We have, for this particular rule Δ,

$$M(t) = \sup_{p_1,p_2} |p_1 - p_2| \, E_\Delta(n(t)|p_1,p_2).$$

It follows from definition (3) that, in general,

(4) $$M(t) \ge M_0(t,\delta).$$

We now turn to a normal approximation, replacing the four Bernoulli sequences by diffusion processes. It will be enough to describe those corresponding to I_1 and I_2, since the others have similar, but independent behaviour. The new information processes $\{x_1(s_1)\}$ and $\{x_2(s_2)\}$ are defined by

$$x_i(s_i) = \mu_i s_i + \tfrac{1}{2} w_i(s_i)$$

where $\mu_i = p_i - \frac{1}{2}$ and $\{w_i(s_i)\}$, $i = 1,2$, are independent standard Wiener processes. It might seem more appropriate to use a diffusion coefficient $p_i(1-p_i) = \frac{1}{4} - \mu_i^2$ here, but we shall be concerned with small values of μ_i. Let us write $y_i = x_i/s_i$ and, for convenience, restrict attention to decision procedures $\Delta = (\Delta(I), \Delta(J))$ based on the sufficient statistics (y_1,y_2,s_1,s_2) at time $s = s_1 + s_2$. The random variables $m(s)$ and $n(s)$ are defined as before and we are concerned with a class of admissible rules similar to C_0. Formally, we say $\Delta \in C_1$ if

(5) $$E_\Delta(m(s))|\mu_1,\mu_2) \le E_\Delta(n(s))|\mu_1,\mu_2)$$

for all μ_1,μ_2 and all values of s in the interval $0 \le s \le t$. Then

(6) $$M_1(t,\delta) = \inf_{\Delta \in C_1} \delta \sup_{|\mu_1 - \mu_2| = \delta} E_\Delta(n(t)|\mu_1,\mu_2).$$

It is intuitively clear that $M_0(t,\delta)$ can be approximated by $M_1(t,\delta)$ when t is large and δ is small. The aim is to find a lower bound on $M_1(t,\delta)$ and then we shall need to rely on the inequality:

$$\text{(7)} \qquad \liminf \; M_0(t,(\omega/t)^{\frac{1}{2}})\, t^{-\frac{1}{2}} \geq \liminf \; M_1(t,(\omega/t)^{\frac{1}{2}})\, t^{-\frac{1}{2}},$$

where ω is a constant. This relation will be assumed without proof.

The next step is to impose a prior distribution on the unknown parameters μ_1, μ_2. Let μ be a random variable with the distribution $N(0,\alpha)$ and suppose that, given μ,

$$(\mu_1,\mu_2) = (\mu+\tfrac{1}{2}\delta,\ \mu-\tfrac{1}{2}\delta) \quad \text{or} \quad (\mu-\tfrac{1}{2}\delta,\ \mu+\tfrac{1}{2}\delta),$$

each with probability $\frac{1}{2}$ Thus, π is confined to the set of points with $|\mu_1-\mu_2| = \delta$ and clearly,

$$\text{(8)} \qquad M_1(t,\delta) \geq \delta \inf_{\Delta \in C_1} E_\Delta(n(t)\,|\,\pi) \,.$$

As we shall see, it is possible to attain the minimum on the right by choosing suitable sampling and allocation rules.

It follows from (5), by integrating with respect to π, that

$$\text{(9)} \qquad E_\Delta(m(s)\,|\,\pi) \leq E_\Delta(n(s)\,|\,\pi)\,.$$

This holds for any procedure $\Delta \in C_1$ and for $0 \leq s \leq t$. We can express

$$\text{(10)} \qquad E_\Delta(n(t)\,|\,\pi) = \int_0^t E_\Delta(u(s)\,|\,\pi)\,ds,$$

where $u(s) = u_\Delta(s)$ represents the probability of allocating the inferior treatment at time s. Consider the problem of minimising this probability with respect to the allocation rule $\Delta(J)$, for a fixed sampling rule $\Delta(I)$. A Bayesian approach to this local problem shows that, given the available data at time s, we must select J_1 or J_2 according to whether the posterior probability associated with the event $[\mu_1 \; \mu_2]$ is greater than or less than $\frac{1}{2}$. The point here is that the posterior distribution of the parameters μ_1,μ_2 depends on the two sampling times s_1,s_2 and also on the histories of the corresponding information processes, but it does not depend explicitly on the sampling rule $\Delta(I)$. In other words, we can evaluate the posterior distribution exactly as if s_1 and s_2 were fixed in advance. This still leaves a substantial calculation, but the Bayes allocation rule can be determined as follows: we must select J_1 or J_2 at time s if and only if $z > 0$, where

$$z = \{\alpha + \frac{1}{8s_2}\} y_1 - \{\alpha + \frac{1}{8s_1}\} y_2.$$

Then it follows, after further calculations, that

$$E_\Delta(u(s)|\pi) \geq E_\Delta[1 - \Phi\{\delta s^{\frac{1}{2}} (\frac{1 + 4\alpha s\eta}{1 + 4\alpha s})^{\frac{1}{2}}\} | \pi], \tag{11}$$

where Φ is the standard normal distribution function and η is the random variable defined by

$$\eta = 4 \frac{m}{s} (1 - \frac{m}{s}), \quad m = \min(s_1, s_2). \tag{12}$$

We now remark that the integrand on the right of (11) is convex and decreasing in m, for values in the range $0 \leq m \leq \frac{1}{2}s$. By using Jensen's inequality, we can replace the random variable $m = m(s)$ by its expectation in (11) and (12). Then, because of (9), we can replace that by $g(s) = E_\Delta(n(s)|\pi)$. By using equation (10) and the modified form of the inequality (11), we obtain the conclusion that

$$(13) \quad \begin{aligned} g(t) &\geq \int_0^t [1 - \Phi\{\delta s^{\frac{1}{2}} (\frac{1 + 4\alpha s\zeta}{1 + 4 s})^{\frac{1}{2}}\}] ds, \\ \zeta &= 4 \frac{g}{s}(1 - \frac{g}{s}). \end{aligned}$$

We have now established that the function g defined by

$$g(s) = E_\Delta(n(s)|\pi) \tag{14}$$

must satisfy the condition (13) when $s = t$, but it is clear from the argument that a similar inequality holds at every point of the interval $0 \leq s \leq t$. These conditions apply for any choice of the decision procedure $\Delta \varepsilon C_1$ and our problem is to minimise the value of $g(t)$. It is a straightforward exercise in analysis to show that

$$h(t) = \min_{\Delta \varepsilon C_1} E(n(t)|\pi) \tag{15}$$

is determined by treating the above inequalities as equations. In other words, the function h is the solution of the differential equation:

(16) $$h' = 1 - \Phi\{\delta s^{\frac{1}{2}}(\frac{1 + 4\alpha s\zeta}{1 + 4\alpha s})^{\frac{1}{2}}\}\ ,\quad \zeta = 4\,\frac{h}{s}(1 - \frac{h}{s}),$$

with $h(0) = 0$. Hence, the relation (8) means that

(17) $$M_1(t,\delta) \geq \delta h(t).$$

Of course, h depends on the constants δ and α, and it remains to choose suitable values for these. In fact, the dependence $h(t) = h(t,\delta,\alpha)$ is quite simple, for reasons closely related to the invariance principle mentioned in the introduction. For any positive constant c,

$$h(t,\delta,\alpha) = c^{-2}h(c^2t,\ c^{-1}\delta,\ c^{-2}\alpha)$$

and, on setting $c = \delta$, we find that (17) becomes

(18) $$M_1(t,\delta) \geq \delta^{-1}h(\delta^2t,\ 1,\ \delta^{-2}\alpha).$$

Finally, we need to combine this with the inequalities (4) and (7). Let $\delta = (\omega/t)^{\frac{1}{2}}$ and $\alpha = a\omega/t$, where ω and a are positive constants. Then we have

$$\liminf M(t)t^{-\frac{1}{2}} \geq \liminf M_1(t,(\omega/t)^{\frac{1}{2}}t^{-\frac{1}{2}} \geq h(\omega,1,a)\omega^{-\frac{1}{2}}.$$

This holds for any choice of ω and a, so that

(19) $$\liminf_{t\to\infty} M(t)/t^{\frac{1}{2}} \geq \sup_{a,\omega} h(\omega,1,a)/\omega^{\frac{1}{2}} = 0.306(8).$$

The supremum here is approached by letting $a \to \infty$ and then by solving the limiting form of equation (16).

I am indebted to my colleague A.H. Craven for computing the solution of this differential equation, which showed that the appropriate choice of ω is about 3.6.

References.

BATHER, J.A. (1980). Randomised allocation of treatments in sequential trials. Adv. Appl.Prob., 12, 174-182.

BATHER, J.A. (1981). Randomised allocation of treatments in sequential experiments. J.R. Statist. Soc. B 43, 265-92.

FABIUS, J. and VAN ZWET, W.R. (1970). Some remarks on the two-armed bandit. Ann. Math. Statist., 41, 1906-1916.

GITTINS, J.C. (1979). Bandit processes and dynamic allocation indices. J.R. Statist. Soc. B 41, 148-177.

ROBBINS, H. (1952). Some aspects of the sequential design of experiments. Bull. Amer. Math. Soc. 58, 527-535.

ROBINSON, D.R. (1982). A comparison of sequential treatment allocation rules: private communication.

VOGEL, W. (1960) (a) A sequential design for the two-armed bandit. (b) An asymptotic minimax theorem for the two-armed bandit problem. Ann. Math. Statist. 31, 430-443, 444-541.

BANDIT PROBLEMS WITH RANDOM DISCOUNTING*

by Donald A. Berry
University of Minnesota

ABSTRACT

One of k independent stochastic processes with unknown characteristics is observed at each of a possibly infinite number of stages. Future stages are discounted: the m*th* observation is weighted by α_m. The α_m are random variables. They may be dependent and their distributions unknown; in such a case one can learn about the character of the discounting as well as about the processes. The objective is to maximize the expected sum of the weighted observations. The decision problem is shown to be equivalent to one with nonrandom discounting in some versions. The important case of geometric discounting arises in a natural way. Other versions are intrinsically more complicated than the nonrandom case. Examples are carried out.

1. INTRODUCTION

One of k stochastic processes is observed at each of a possibly infinite number of stages. Selecting a process (or *arm*) to observe is called a *pull*. The arm pulled at any stage can depend on the pulls and resulting observations at all previous stages.

A *strategy* is a function that, for each finite history of pulls and observations, assigns an arm to be pulled next. To stress dependence on the strategy, τ_m will denote the observation at stage m when following strategy τ. If τ specifies arm j at stage m then $\tau_m = X_{jm}$. (For notational convenience it is assumed that all k processes are ongoing though only one can be observed at a time.)

Assume for fixed j that the X_{jm}, m = 1, 2, ..., are identically distributed and independent given a common parameter θ_j. At least one of the θ_j is unknown, for otherwise the problem would be trivial. The parameters are themselves random variables with given "prior" probability distributions. So if θ_j is unknown, variables X_{jm}, m = 1, 2, ..., are exchangeable rather than independent--learning is possible. The information available about arm j at any time is contained in the current probability distribution on θ_j.

Such decision problems are sometimes called "bandits" in analogy with choosing whether or not to play a slot machine--colloquially called a "one-armed bandit." Most of the bandit literature treats one of two

*Research supported by the National Science Foundation under Grant No. MCS81-02477.

objectives:

(i) *Finite horizon*: for some fixed n, the expected sum of the first n observations is to be maximized.

(ii) *Geometric discounting*: the m*th* observation is weighed by a factor α^m, $0 < \alpha < 1$, and the expected weighted sum over the infinite horizon is to be maximized.

Historically important papers concerning these objectives are, respectively, (Bradt, Johnson and Karlin 1956) and (Bellman 1956)--both papers deal with Bernoulli processes. Very recent papers by participants in this conference, again respectively, are (Bather 1981) and (Gittins 1979).

A general discounting approach, which includes objectives (i) and (ii), is taken in (Berry and Fristedt 1979)--referred to henceforth as BF79. The m*th* observation is weighed by a factor α_m and the expected weighted sum over the infinite horizon is to be maximized. So a strategy is *optimal* if it maximizes expected payoff:

$$(1.1) \qquad W(\tau) = E\,\Sigma_{m=1}^{\infty}\,\alpha_m \tau_m .$$

When the discount factors α_m are known constants, which is assumed throughout this section, (1.1) becomes

$$(1.2) \qquad W(\tau) = \Sigma_{m=1}^{\infty}\,\alpha_m E\tau_m .$$

Assume $\alpha_m \geq 0$ for all m and $\Sigma_1^{\infty}\alpha_m < \infty$; $A = (\alpha_1, \alpha_2, \ldots)$ is called a *discount sequence*. That (ii) is a special case is obvious; for (i) take $\alpha_1 = \ldots = \alpha_n = 1$ and $\alpha_{n+1} = \ldots = 0$.

Because the language is so appealing, the arm specified at the first stage by an optimal strategy is called an "optimal arm."

An easy example may underscore some critical issues.

Example 1.1. Suppose $A = (1, 1, 0, \ldots)$; that is, (i) applies with $n = 2$. Each $\{X_{jm}: m = 1, 2, \ldots\}$ is a Bernoulli process with $\theta_j = P(X_{jm} = 1)$; assume the k processes are independent. There are k^3 essentially different strategies. This number can be reduced to k^2 by applying the stay-on-a-winner rule (Berry 1972): If an optimal arm is pulled at any stage and yields a success, then it is optimal at the next stage as well. Label the arms so that $E\theta_1 \geq \ldots \geq E\theta_k$. We need only consider strategies that use arm 1 after a failure on the first pull of any arm other than 1. For, by Cauchy-Schwarz,

$$P(X_{j2} = 1 | X_{j1} = 0) = \frac{E\theta_j - E\theta_j^2}{1 - E\theta_j} \leq \frac{E\theta_j(1 - E\theta_j)}{1 - E\theta_j} = E\theta_j \leq E\theta_1 .$$

There are two possibilities--arm 1 and arm 2--when arm 1 is used initially and fails.

There are k + 1 strategies to consider: $\tau^0, \tau^1, \ldots, \tau^k$. In an evident notation, and using independence,

$$W(\tau^0) = 2E\theta_1,$$

$$W(\tau^1) = E\theta_1 + E\theta_1^2 + (1 - E\theta_1)E\theta_2,$$

$$W(\tau^j) = E\theta_j + E\theta_j^2 + (1 - E\theta_j)E\theta_1,$$

for j = 2, ..., k. And τ^j is optimal if its expected payoff is greatest.□

A famous result of Gittins and Jones (1974) says, in the case of geometric discounting, that optimal strategies can be found by comparing each arm in turn with known arms. This implies that the preference for arms is transitive: arm 1 preferred to arm 3 and arm 2 preferred to arm 1 implies arm 2 preferred to arm 3. The setting of the previous example gives a counterexample when the discount sequence is not geometric. The following is a continuation of that example and is essentially the same as one communicated to me by Bert Fristedt.

Example 1.2 (continuation of Example 1.1). Suppose $E\theta_1 = 0.5$, $E\theta_1^2 = 0.25$ (so θ_1 is known to be 0.5), $E\theta_2 = 0.475$, $E\theta_2^2 = 0.27625$ (e.g., θ_2 is either 0.7 or 0.25 with equal probabilities), $E\theta_3 = E\theta_3^2 = 0.33$ (so θ_3 is 0 or 1 with probabilities 0.67 and 0.33). Then $W(\tau^0) = 1$, $W(\tau^1) = 0.9875$, $W(\tau^2) = 1.01375$, and $W(\tau^3) = 0.995$. So τ^2 is optimal.

Now consider the three possible two-armed bandits: arms 1 and 3, arms 1 and 2, arms 2 and 3. By virtue of the previous calculations, arm 1 is optimal in the first and arm 2 in the second. But, as the following calculations show, arm 3 is optimal in the third! The optimal worth pulling arm 2 first is

$$E\theta_2 + E\theta_2^2 + (1 - E\theta_2)\max\left\{\frac{E\theta_2 - E\theta_2^2}{1 - E\theta_2},\ E\theta_3\right\}$$

$$= 0.475 + 0.27625 + 0.525\max\{0.3786,\ 0.33\}$$

$$= 2(0.475) = 0.95.$$

The optimal worth pulling arm 3 first is

$$E\theta_3 + E\theta_3^2 + (1 - E\theta_3)\max\left\{\frac{E\theta_3 - E\theta_3^2}{1 - E\theta_3},\ E\theta_2\right\}$$

$$= 0.33 + 0.33 + 0.67\max\{0,\ 0.475\} = 0.97825.$$ □

The case k = 2 is considered in BF79; the characteristics of one arm, say arm 1 for definiteness, are unknown and those of arm 2 are known. So the information concerning arm 1 changes as it is pulled, but that of arm 2 does not. It is well-known in this case for both discount sequences (i) and (ii) that there exists an optimal strategy with the following characteristic: once arm 2 is selected it is thenceforth used exclusively and indefinitely. Such problems are stopping problems: one need only decide when to stop experimenting with arm 1. BF79 shows there are always optimal strategies with this characteristic if the discount sequence (assumed to be monotonic) is *regular*. Conversely, if it is not regular then there is a distribution on θ_1 for which no optimal strategies have this characteristic (cf. Example 1.3).

Definition 1.1. A discount sequence $A = (\alpha_1, \alpha_2, \ldots)$ is *regular* if, for each m,

$$\gamma_m \gamma_{m+2} \leq \gamma_{m+1}^2$$

where $\gamma_r = \Sigma_{i=r}^{\infty} \alpha_i$.

(Replacing each γ_m with α_m in this definition gives a stronger condition, called *superregularity* in (Berry and Fristedt 1982).)

The following are examples.

Regular:

(iii) $(1, \ldots, 1, \alpha, \alpha^2, \ldots)$, $0 < \alpha < 1$

(iv) (4, 4, 3, 3, 2, 2, 1, 1, 0, ...)

(v) (2, 1, 1, 0, ...)

Not regular:

(vi) (2, 1, 1, 1, 0, ...)

(vii) (4, 1, 1, 0, ...)

(viii) $(1/2, 5/16, \ldots, (1/2)(3/4)^m + (1/2)(1/4)^m, \ldots)$

That sequence (v) is regular follows from the regularity of (iv); it is listed for easy comparison with (vi).

Sequence (viii) is the average of two geometrics, which, of course, are themselves regular. But geometrics are barely regular: $\gamma_{m+1}^2 = \gamma_m \gamma_{m+2}$ for all m. So the slightest tampering destroys regularity. In particular, means of nondegenerate mixtures of geometrics are never regular, as the following calculation shows. Consider the sequence $(EV, EV^2, EV^3, \ldots)$ where V is a random variable on [0,1]. Then, for m = 1, 2, ...,

$$\gamma_m = E\left(\frac{V^m}{1 - V}\right).$$

We have

$$\gamma_2^2 - \gamma_1\gamma_3 = E^2\left(\frac{V^2}{1 - V}\right) - E\left(\frac{V^2}{1 - V} + V\right)E\left(\frac{V^2}{1 - V} - V\right)$$
$$= E\left(\frac{VEV^2 - V^2EV}{1 - V}\right).$$

The function $(xEV^2 - x^2EV)/(1 - x)$ is concave in x on [0,1]--strictly concave unless V = 0 or V = 1 with probability one. Therefore, Jensen's inequality applies to show that

$$\gamma_2^2 - \gamma_1\gamma_3 \le 0$$

with strict inequality provided V is not concentrated at one point.

Example 1.3. Suppose k = 2. As in Example 1.1, the processes are Bernoulli with, for j = 1, 2, $\theta_j = P(X_{jm} = 1)$. Suppose θ_2 is known and θ_1 is either 0 or 1 with probabilities 1/2 each. This assumption makes the problem relatively easy because a single observation on arm 1 reveals θ_1. If the discount sequence is regular then the problem is trivial because only two strategies need be considered. Namely, τ': pull arm 1, if $\tau_1' = 1$ (success) pull arm 1 forever and if $\tau_1' = 0$ (failure) pull arm 2 forever; and τ'': pull arm 2 forever.

Consider discount sequence (viii). Since it is not regular we must allow for switches to arm 1 from arm 2. The set of optimal strategies depends on θ_2. A complete description follows. If $\theta_2 \in [0,\lambda_0]$ then use τ'. If $\theta_2 \in [\lambda_0,\lambda_1]$ then pull arm 2 and follow with τ'. If $\theta_2 \in [\lambda_1,\lambda_2]$ then pull arm 2 twice and follow with τ'. In general, if $\theta_2 \in [\lambda_{s-1},\lambda_s]$ it is optimal to pull arm 2 s times and then to use τ'. Here, $0.7273 \doteq 8/11 = \lambda_0 < \lambda_1 < \lambda_2 < \ldots < \lambda_\infty = \lim \lambda_s = 4/5$. If $\theta_2 \in [4/5,1]$ then τ'' is optimal. The first few values of λ_s are, to four decimals, 0.7273, 0.7692, 0.7887, 0.7961, 0.7987, 0.7996, 0.7999. Even though the structure is otherwise simple, the fact that the discount sequence is not regular makes the solution complicated.□

The possibility that the discount factors are unknown is introduced in the next section. Allowing for randomness in the discount sequence is natural enough, but it seems not to be considered in the literature--not in the bandit literature anyway. Two versions are considered depending on whether the discount factors are observable. When they are not, or when they must be ignored, the problem is shown to be equivalent to one with nonrandom discounting. When they are, it sometimes reduces to a nonrandom problem and sometimes does not.

2. PRELIMINARIES

Suppose the discount sequence is not completely known. In economics, for example, the inflation rates in future years would not be known. In a medical trial the size of the patient pool may itself be random. Or, a new arm may be discovered--one that is obviously better than the arms in the trial. This would likely end the trial prematurely; the discount factors become 0 from some stage on, and that stage is random.

One way to allow a discount sequence to be random is to place a measure on the space of nonrandom sequences. A random discount sequence is the corresponding mixture of nonrandom ones. However, specifying a measure with a large support is difficult. The bulk of this article takes a narrower approach, but one that is natural and seems easy to apply. Mixtures will be discussed again in Section 5.

Let $U_1, U_2, \ldots$ be nonnegative random variables. Set $\alpha_1 = U_1$ and for $m = 2, 3, \ldots,$ recursively define

$$\alpha_m = \alpha_{m-1} U_m.$$

The distribution measures of $U_1, U_2, \ldots,$ call them $F_1, F_2, \ldots,$ may themselves be unknown. Given $F_1, F_2, \ldots,$ variables $U_1, U_2, \ldots$ are assumed to be independent. However, if the F_i are dependent random distributions then the U_i are not generally independent.

It will be assumed throughout that the U_i are independent of the X_{jm}.

There is now some ambiguity in the use of the term "strategy." This will be resolved momentarily. In any case definition (1.1) of expected payoff of a strategy τ continues to apply with

$$\alpha_m = \Pi_1^m U_i .$$

The expectation in (1.1) is now with respect to the distribution of the U_i as well as that of the τ_m.

We shall consider two sets of ground rules:

Version 1. The random variables U_i are not observable. So while the τ_m are observed, the discounted payoff at stage m, $\alpha_m \tau_m$, is not. The set of available strategies in this version, call it T_1, is as defined in Section 1 for the nonrandom case.

Version 2. The random variables U_i are observable. The decision at stage $m+1$ can depend on $(U_1, \ldots, U_m)$ as well as on τ and $(\tau_1, \ldots, \tau_m)$. Let T_2 denote the corresponding set of available strategies.

A third possibility--one not considered here--is that the product

$\alpha_m\tau_m$ is observed at stage m, but not α_m and τ_m individually.

Version 2 seems more realistic than Version 1. But one can imagine circumstances in which a strategy can be programmed to depend only on the results of the pulls. Strategies in T_1 are simpler than typical strategies in T_2. Actually, each $\tau \in T_1$ has a version in T_2: there is a strategy in T_2 which duplicates the decisions specified by any $\tau \in T_1$. Therefore, Version 1 provides a bound for Version 2: the maximal expected payoff in Version 2 is no smaller than in Version 1. Typically, it is greater. But, as will be seen, there are numerous circumstances in which they are equal, when the ability to observe the U_i provides no advantage.

3. VERSION 1: NONOBSERVABLE DISCOUNT FACTORS

For all $\tau \in T_1$, $(\tau_1, \tau_2, \ldots)$ is independent of $(U_1, U_2, \ldots)$. Therefore, (1.1) becomes

$$(3.1) \qquad W(\tau) = \Sigma_1^\infty E\alpha_m E\tau_m$$

for all $\tau \in T_1$, where

$$(3.2) \qquad E\alpha_m = E\,\Pi_1^m U_i.$$

So (1.2) applies with α_m replaced by $E\alpha_m$. And the problem considered here is no more general than that considered in BF79 (except that the number of arms is now arbitrary and the possibility $E\alpha_{m+1} > E\alpha_m$ is not ruled out).

In the special case in which the U_i are independent, (3.2) becomes

$$(3.3) \qquad E\alpha_m = \Pi_{i=1}^m EU_i.$$

Example 3.1. Suppose the U_1 are independent with

$$EU_1 = \ldots = EU_n = 1,\ EU_{n+1} = \ldots = 0$$

(F_{n+1} concentrates its mass at 0 and the F_i for $i > n + 1$ are immaterial). Then the discount sequence relevant for choosing a strategy is (i), finite horizon: EA = (1, 1, ..., 1, 0, ...). This is not to say the choice is easy. But the same backward induction is available for finding optimal strategies as in the usual, nonrandom finite horizon setting.□

Example 3.2. Suppose the U_i are independent with $EU_i = \alpha$ for i = 1, 2, ...; α is known and $0 < \alpha < 1$. It may be, for example, that the trial terminates at stage m with conditional probability $1 - U_m$. Then $E\alpha_m = \alpha^m$ and the problem is the same as (ii), geometric discounting. In particular, the results of (Gittins 1979) apply.□

The nonrandom discount sequences in the previous two examples are regular. The resultant sequence in the next example is not regular. It will be referred to again in Example 4.1.

Example 3.3. Discount sequence (viii) considered in Example 1.3 is (1/2, 5/16, 7/32, ...). This can arise as the mean of $A = \{\alpha_m\}$ in a number of ways. For example, the U_i may be independent (so (3.3) applies) with

$$EU_i = (3^i + 1)/[4(3^{i-1} + 1)]$$

for i = 2, 3, ... Or, $P(F_1 = F_2 = \ldots = F) = 1$ where F is an equal mixture of two one-point distributions; one at 3/4 and one at 1/4. In the latter interpretation $P(U_1 = U_2 = \ldots = 3/4) = P(U_1 = U_2 = \ldots = 1/4) = 1/2$. This is consistent with viewing A as the average of two geometrics. Regardless of how the sequence arises, an optimal strategy is as given in a nonrandom setting with discount sequence EA = (1/2, 5/16, 7/32, ...); for a special case see Example 1.3.□

Example 3.4. Suppose $F_1 = F_2 = \ldots = F$ where $F(\{1\}) = q = 1 - F(\{0\})$; q is unknown and has a uniform distribution on (0,1). This seems to be a harmless assumption. However,

$$E\alpha_m = P(U_1 = \ldots = U_m = 1) = \int_0^1 q^m dq = \frac{1}{m+1}.$$

So $\Sigma E\alpha_m = \infty$ and EA is not a discount sequence. (If $\Sigma E\alpha_m = \infty$ were allowed Then EA would not be regular. For a sequence with infinite sum one would eschew immediate gain and sample only to obtain information that might help in the long run. Optimal strategies would be similar to Kelly's (1981) "least-failures rule.")□

4. VERSION 2: OBSERVABLE DISCOUNT FACTORS

Strategies in T_1 do not depend on the U_i. Strategies in T_2 depend on the U_i as well as the observed X_{jm}. This section treats the latter possibility.

There is an important distinction in Version 2 between independent and dependent U_i. These cases are considered separately.

4.1. Independent U_i

Suppose for i = 1, 2, ... that F_i is a random distribution with measure μ_i on the space of distributions. F_i is known if μ_i is a one-point measure. For the purposes of this section assume the F_i are independent. Then so are the U_i.

In making a decision at stage m + 1, $U_1, \ldots, U_m$ are known. Since the U_i are independent the conditional distribution of $U_{m+1}, U_{m+2}, \ldots$ given $U_1, \ldots, U_m$ is the same as the unconditional. Assume $U_1 = u_1, \ldots, U_m = u_m$. The decision problem at stage m + 1 is to maximize

$$\Sigma_{r=m+1}^{\infty} (\Pi_{i=1}^{m} u_i)(\Pi_{i=m+1}^{r} EU_i)E\tau_r,$$

where the expectations are conditional on the U_i. This can be written

$$K\, \Sigma_{r=m+1}^{\infty} (\Pi_{i=1}^{r} EU_i)E\tau_r,$$

where

$$K = \Pi_{i=1}^{m}(u_i/EU_i).$$

Decision problems with proportional discount sequences are equivalent. An optimal selection at stage j + 1 can be made without observing the U_i; equivalently, each U_i can be assumed equal to its mean. Therefore, an optimal strategy in T_2 can be found by maximizing (3.1). And (3.3) applied where the mean of U_i can be expressed as

$$EU_i = \int E(U_i|F_i)\mu_i(dF_i) .$$

So when the F_i are independent the problem is the same whether or not the discount factors are observable. And in turn both random discounting versions are equivalent to nonrandom discounting. Moreover, the expected payoff of any strategy is the same in all three cases. Of course, the expected payoff of the continuation of a strategy changes depending on the u_i.

Examples 3.1, 3.2 and 3.3 apply also for the case considered here. Take Example 3.1. The mean of the discount sequence relevant at stage 2, given $U_1 = u_1$, is u_1 times the (n - 1)-horizon: (1, 1, ..., 1, 0, ...). Each new stage gives a problem identical with the corresponding one in Example 3.1.

4.2. Dependent U_i

Some additional notation is helpful for this case. The ideas apply generally but for convenience the development is restricted to the Bernoulli case: every pull results in a 0 or a 1. The *jth* arm gives 1 with probability θ_j.

The (initial) random discount sequence is

$$A = (U_1, U_1U_2, U_1U_2U_3, \ldots).$$

At stage 2, after observing U_1, the relevant discount sequence is

$$\begin{aligned}(A^{(1)}|U_1) &= (U_1U_2|U_1, U_1U_2U_3|U_1, \ldots)\\ &= U_1(U_2|U_1, U_2U_3|U_1, \ldots);\end{aligned}$$

this and subsequent notation is consistent with BF79.

Let G denote the initial joint distribution of $(\theta_1, \ldots, \theta_k)$. If arm j is pulled and results in success, $X_{j1} = 1$, then G is changed via Bayes theorem to $\sigma_j G$, say. Similarly, a failure on arm j changes G to $\varphi_j G$.

Let V_j denote the expected payoff of pulling arm j initially and then following an optimal strategy (in T_2). Define

$$V = \max\{V_1, \ldots, V_k\}.$$

The relevant standard functional equations are

$$\begin{aligned}(4.2)\qquad V_j(A,G) &= E\theta_1 EU_1 + E\theta_1 E[U_1 V((A^{(1)}|U_1),\sigma_j G)]\\ &\quad + (1 - E\theta_1)E[U_1 V((A^{(1)}\ U_1),\varphi_j G)],\end{aligned}$$

for j = 1, ..., k. The problem can be solved, or at least the solution approximated, by repeated application of (4.2). But the calculations can be forbidding. In particular, the posterior distribution of $(U_{m+1}, U_{m+2}, \ldots)$ given $U_1, \ldots, U_m$ can be arbitrarily difficult unless a simple structure is imposed.

To make the calculations manageable, assume the unknown F_i have a special kind of dependence: for all i, $F_i = F$ which is a random distribution with measure μ. When a discount factor α_m--and therefore U_m--is observed, the current measure of F is updated. Updating is easiest if F is known up to some real-valued parameter η. For then Bayes theorem applies to modify a prior distribution on η.

A useful alternate approach due to Ferguson (1973) is to give F a "Dirichlet process prior." For each real u, F(u) has a beta distribution with parameters $MF_0(u)$ and $M(1 - F_0(u))$; F_0 is the prior mean of F and M is a measure of prior precision. After observing $U_1 = u_1, \ldots, U_m = u_m$, the posterior of F is also a Dirichlet process. The new M is M + m and MF_0 becomes $MF_0 + \Sigma_1^m I_{u_i}$; here, $I_x(u) = 1$ if $u \leq x$ and 0 otherwise. This approach has promise for two reasons: (1) As is clear from the above comments, calculations are manageable. (2) The support of a Dirichlet process (in the topology of pointwise convergence) contains all probability measures absolutely continuous with respect to F_0 (Ferguson 1973).

Neither of the above-mentioned possibilities for updating the distribution of F are carried forward in the present paper. (I plan more work on this problem.) Instead, an example is given in which updating is quite simple.

Example 4.1. Consider the setting of Example 1.3: there are two Bernoulli arms, θ_2 is known, and θ_1 is either 0 or 1, with equal probabilities under G. Distribution F is unknown; it is one of two one-point distributions with equal probabilities, one point is 3/4 and the other is 1/4. Therefore the U_i are either all 3/4 or all 1/4; which one will be revealed at the first stage.

In Version 1 (see Example 3.3) the relevant discount sequence, EA = (1/2, 5/16, 7/32, ...), is not regular. When F is unknown regularity of EA is not a consideration. However, F becomes known after stage 1. And, for u = 3/4 or u = 1/4,

$$(A^{(1)} | U_i = u) = u(u, u^2, u^3, \ldots),$$

with probability one. Since both these sequences are geometric, and therefore regular, the number of strategies in T_2 that must be considered is sharply reduced.

A further reduction is possible. If the U_i were known in advance, Example 4.4 of BF79 applies in a trivial way. The "break-even value" of θ_2 when $U_1 = 3/4$ is $\theta_2 = 4/5$; when $U_1 = 1/4$ it is 4/7. The number of strategies to consider can be reduced to three: τ'--pull arm 1, pulling it indefinitely if it is successful and switching to arm 2 (permanently) otherwise; τ''--pull arm 2 indefinitely; τ'''--pull arm 2, then follow τ' if $U_1 = 3/4$ and τ'' if $U_1 = 1/4$. Easy calculations show: $W(\tau') = 7\theta_2/12 + 5/6$, $W(\tau'') = 5\theta_2/3$, $W(\tau''') = 185\theta_2/192 + 9/16$. So $V_1(A,G) = W(\tau')$ and $V_2(A,G) = \max\{W(\tau''), W(\tau''')\}$ and $V(A,G) = \max\{W(\tau'), W(\tau''), W(\tau''')\}$. All optimal strategies are given as follows: τ' for $\theta_2 \leq 52/73 \doteq 0.7123$,

τ''' for $52/73 \leq \theta_2 \leq 4/5$, and τ'' for $\theta_2 \geq 4/5$.

This solution should be compared with that of Example 1.3. The interested reader can check that

$$\sup_{\tau \in T_1} W(\tau) \leq \sup_{\tau \in T_2} W(\tau)$$

with strict inequality if and only if $52/73 < \theta_2 < 4/5$.

In this example, not only is Version 2 an improvement over Version 1, but the analysis is simpler.□

5. MIXTURES

As indicated in Section 2, a more general way of introducing random discount sequences is to mix nonrandom sequences. In Version 1, nonobservable discount factors, the problem reduces to one with a nonrandom discount sequence. The reasons given in Section 3 also apply for mixtures. The corresponding nonrandom sequence is simply the mean of the random sequence.

Consider Version 2, observable discount factors. After stage m the mixing distribution is updated via Bayes theorem in a very simple way. Suppose $\alpha_1', \ldots, \alpha_m'$ are known to be the first m discount factors. The total posterior probability of those sequences which disagree with $(\alpha_1', \ldots, \alpha_m')$ in at least one of the first m positions is 0. And the posterior measure of those not ruled out is proportional to the initial measure.

For example, suppose all the sequences in the support of the initial distribution have distinct first factors. Then the true discount sequence will be revealed at stage 1. Learning takes place quickly, but this brings out a difficulty in applying the mixture approach. If one has not been sufficiently careful assigning the initial distribution then every discount sequence may soon be ruled out! And it is difficult to assign a measure rich enough to avoid this problem. In the approach of previous sections, one worries about randomness in a discount sequence on a day-to-day, or stage-to-stage, basis. With mixtures one continually worries about an eternity of randomness.

Example 5.1. Suppose every sequence in the support of the initial measure is of the form (i), finite horizon: (1, 1, ..., 1, 0, ...), differing only in the length of the horizon. In this rather special circumstance, observations of the discount factors can be ignored: Version 2 = Version 1. For, the decision maker can always act as though the discount factor "1" was just observed; if it really was a "0" then the remaining actions are of no consequence.

Every nonrandom discount sequence can be expressed as the mean of a mixture of finite horizons. Suppose, for example, the initial probability of (0, 0, ...) is $1 - \alpha$, where α is known and $0 < \alpha < 1$, of (1, 0, ...) is $(1 - \alpha)\alpha$, of (1, 1, 0, ...) is $(1 - \alpha)\alpha^2$, etc. Then the mean of this mixture is the geometric sequence, (ii): $(\alpha, \alpha^2, \ldots)$. So in this setting, optimal strategies in Version 1 are also optimal in Version 2. Moreover, they can be found from the nonrandom geometric discounting case.□

6. CONCLUSION

When discount factors ΠU_i are random but cannot be observed, the problem is identical with a particular nonrandom problem.

When such discount factors can be observed and the U_i are independent random variables, then again the problem reduces to one that is nonrandom. If, in addition, the U_i are identically distributed then the appropriate discount sequence is geometric. I regard this to be realistic and an argument for geometric discounting rather than, for example, finite horizon.

The problem does not reduce to one that is nonrandom when the U_i are dependent and learning about the future U_i is possible. The set of available strategies is larger in this case. However, the task of finding an optimal strategy can be easier.

ACKNOWLEDGEMENT

I want to thank Bert Fristedt, John Bather, Ronald Christensen, Alfonso Novales, and David Polansky for helpful discussions.

REFERENCES

Bather, J. A. (1981). Randomized allocations of treatments in sequential experiments (with discussion). J. R. Statist. Soc. B 43:265-292.

Bellman, R. (1956). A problem in the sequential design of experiments. Sankhya A 16:221-229.

Berry, D. A. (1972). A Bernoulli two-armed bandit. Ann. Math. Statist. 43:871-897.

Berry, D. A., and Fristedt, B. E. (1979). (Called BF79 in text.) Bernoulli one-armed bandits--Arbitrary discount sequences. Ann. Statist. 7:1086-1105.

Berry, D. A., and Fristedt, B. E. (1982). Maximizing the length of a success run for many-armed bandits. Stochastic Processes and Their Applications (to appear).

Bradt, R. N., Johnson, S. M., and Karlin, S. (1956). On sequential designs for maximizing the sum of n observations. Ann. Math. Statist. 27:1060-1070.

Ferguson, T. S. (1973). A Bayesian analysis of some nonparametric problems. Ann. Statist. 1:209-230.

Gittins, J. C. (1979). Bandit processes and dynamic allocation indices (with discussion). J. Roy. Statist. Soc. B 41:148-177.

Gittins, J. C., and Jones D. (1971). A dynamic allocation index for the sequential design of experiments. Progress in Statistics (ed. by J. Gani et al.), pp. 241-266, North-Holland, Amsterdam.

Kelly, F. P. (1981). Multi-armed bandits with discount factor near one: the Bernoulli case. Ann. Statist. 9:987-1001.

STOCHASTIC APPROXIMATION ON A BOUNDED CONVEX SET

Václav Dupač and Tomáš Fiala
Charles University, Prague

1. Introduction and general assumptions

Let us consider the multidimensional Robbins-Monro stochastic approximation procedure restricted to a bounded convex set by projections. This is defined recursively by

$$X(0) = x_0,$$

$$(1)\qquad X^*(t) = X(t) + a(t+1)(R(X(t)) + G(t+1, X(t), \omega)),$$

$$X(t+1) = \pi_C(X^*(t)), \qquad t \in \mathbb{N}_0,$$

where

(i) $x_0 \in \mathbb{R}^n$ is an arbitrary initial value;

(ii) $C \subset \mathbb{R}^n$ is a bounded closed convex set with nonempty interior C^o, π_C denoting the projection onto C;

(iii) R is a Borel measurable mapping $C \to \mathbb{R}^n$ with a unique $\theta \in C$ such that $R(\theta)=0$; we assume that this θ lies in C^o;

(iv) $(G(t,x,\omega),\ x \in C)_{t=1}^{\infty}$ is a sequence of random functions which are $\mathcal{G}_t \times \mathcal{B}$ measurable, $\mathcal{G}_t = \sigma\{G(s,x,\omega),\ 1 \leq s \leq t,\ x \in C\}$, $\mathcal{B}$ is the σ-algebra of Borel subsets of C; $G(t,x,\omega)$ independent of $\mathcal{G}_{t-1}$, $EG(t,x,\omega)=0$, $\forall t \in \mathbb{N}$, $x \in C$;

(v) $(a(t))_{t=1}^{\infty}$ is a sequence of positive constants such that

$$\sum_{t=1}^{\infty} a(t) = \infty, \qquad \sum_{t=1}^{\infty} a^2(t) < \infty.$$

Further assume

(vi) $\sup_{x \in C \cap U_\varepsilon^c(\theta)} \langle R(x),\ x-\theta \rangle < 0,\ \forall \varepsilon > 0$,
where $U_\varepsilon(\theta)$ denotes the ε-neighbourhood of θ;

(vii) $|R(x)|^2 + E|G(t,x,\omega)|^2 \leq K$, $\forall t \in \mathbb{N}$, $x \in C$ and some $K > 0$.

Then we have $X(t) \to \theta$ a.s.

Instead of projections, we may use the penalty function method as an alternative approach. Results in both directions can be found in Kushner, Clark (1978); see also Nevel´son, Has´minskiĭ (1972), Chapter 7, from where the above stated theorem immediately follows.

In some situations, however, calculating projections might be uncomfortable; some objections may be raised also against the penalty function method : for instance, points outside C might be inaccessible for observations.

For such situations, we propose and investigate another modification of the Robbins-Monro procedure, defined by

$$X(0) = x_0,$$

$$(2)\quad X(t+1) = \begin{cases} X^*(t) & \text{for } X^*(t) \in C, \\ X(t) & \text{for } X^*(t) \notin C, \end{cases} \qquad t \in \mathbb{N}_0,$$

where $X^*(t)$ has the same meaning as in (1).

No projections are now calculated; the new approximation is accepted only if it lies in C, otherwise we stick to the preceding one and try to find an admissible new approximation again.

It is easy to see that this time some additional assumptions are needed to avoid situations like the one, where $x+R(x)+G(t,x,\omega)$ lies outside C for some $x \in \partial C$ and all t with probability 1 and hence typically also

$$(3)\qquad x^* = x + a(t+1)(R(x) + G(t+1,x,\omega))$$

does, which entails that the approximation sequence having once reached the point x will stay there for ever. Even if the probabilities of $x+R(x)+G(t,x,\omega)$ falling into C are all positive, there are still situations to be avoided. In fact, if we look at the problem more carefully, we could see, that what is now actually needed is not (vi), but (vi) with $R(x)$ replaced by

$$(4)\qquad R^*(t,x) = E((R(x)+G(t+1,x,\omega))1_{[x^* \in C]})$$

(and with the supremum taken also over all t). For $x \in C^o$ and $t \to \infty$, $[x^* \in C]$ tends to certainty and hence $R^*(t,x)$ tends to $R(x)$. However, for $x \in \partial C$ and all t (as well as for fixed t and x close to ∂C), $R^*(t,x)$ may be substantially different from $R(x)$ and hence the modified assumption (vi) may be violated even if the original one is fulfilled.

Both types of situations can be avoided by making proper additional assumptions, or alternatively, and this is probably more useful, by modifying the procedure at the boundary ∂C and close to it, viz. by introducing an additional, artifidal noise component to the observations.

2. Convergence theorems for the proposed procedure

Let us start with formulating additional assumptions ensuring convergence of the procedure (2). They look reasonable, although rather special; some weakening of them is possible, anyway.

Theorem 1. Consider the procedure (2). Retain all the assumptions (i)-(vii) of Section 1 (except that π_C is not made use of now). Assume in addition:

(viii) $C=\{y:F(y)\leq 0\}$, $\partial C=\{y:F(y)=0\}$, $\overset{\circ}{C}=\{y:F(y)<0\}$, for some function $F:\mathbb{R}^n\to\mathbb{R}$ with continuous second order derivatives and positive definite Hessian $\ddot{F}(y)$, $\forall y\in\mathbb{R}^n$;

(ix) R is Lipschitz continuous in a (relative in C) neighbourhood of ∂C;

(x) the distribution of $G(t,x,\omega)$ is multidimensional normal, not depending on t; the covariance matrix $B_x=E(G(1,x,\omega)G^T(1,x,\omega))$ is positive definite on ∂C; the matrix $B_x^{1/2}$ is Lipschitz continuous in a (relative in C) neighbourhood of ∂C;

(xi) $\sum_{t=1}^{\infty} a^{2-\varepsilon}(t)<\infty$ for some $\varepsilon>0$;

(xii) $\sup_{x\in\partial C}\langle -B_x\dot{F}(x),x-\theta\rangle<0$.

Then we have $X(t)\to\theta$ a.s.

Remark. The assumption (xii) is satisfied especially for $B_x = \sigma^2 I$ ($\sigma^2>0$ a scalar, I the identity matrix), i.e. if the components of $G(1,x,\omega)$ are independent and identically $N(0,\sigma^2)$ distributed. The same is true, if σ^2 varies with x, provided that (x) is fulfilled. The explanation is that for $x\in\partial G$, the angle between $\theta-x$ and the inward pointing gradient $-\dot{F}(x)$ is always less than $\pi/2$.

We can get rid of the assumption (xii) by introducing an additional, artificial noise (with independent and homoscedastic components) to observations at the boundary ∂C or close to it:

Theorem 2. Consider the procedure (2), with $X^*(t)$ redefined by

(5) $$X^*(t) = X(t) + +a(t+1)(R(X(t))+G(t+1,X(t),\omega)+H(t+1,\omega)1_{[X(t)\in C\cap U(\partial C)]}) ,$$

where $H(t,\omega)$, $t\in\mathbb{N}$, are independent, also of the $G(t,x,\omega)$'s, all normally $N(0,bI)$ distributed random vectors, $b>0$, and $U(\partial C)$ is an (arbitrarily chosen) neighbourhood of ∂C. Assume (i)-(xi). If

(xiii) $b>\sup_{x\in\partial C}\langle -B_x\dot{F}(x),x-\theta\rangle/|\langle -\dot{F}(x),x-\theta\rangle|$,

then we have $X(t)\to\theta$ a.s.

3. The proofs of Theorems 1 and 2

$(X(t))_{t=1}^{\infty}$ defined by (2) is a Markov sequence. Let us investigate its generating operator

$$LV(t,x) = E(V(t+1,X(t+1)) - V(t,X(t)) \mid X(t) = x) =$$
$$= E(V(t+1, x+a(t+1)(R(x)+G(t+1,x,\omega))1_{[x^*\in C]}) - V(t,x),$$

applied to $V(t,x) = |x-\theta|^2/2$:

(6) $$L(|x-\theta|^2/2) = a(t+1)\langle R^*(t,x), x-\theta\rangle +$$
$$+ (1/2)a^2(t+1)E(|R(x)+G(t+1,x,\omega)|^2 1_{[x^*\in C]}),$$

where $R^*(t,x)$ is defined by (4). The second term on the right of (6) is bounded by $(1/2)Ka^2(t+1)$, according to (vii); let us investigate the first term. We start with the inequality

(7) $$P(x^*\notin C) \leqq P(|G(1,x,\omega)| > \frac{d(x,\partial C)}{a(t+1)} - R_0),$$

where $d(x,\partial C)$ is the distance of x to ∂C and $R_0 = \sup_{x\in C}|R(x)|$; the inequality (7) is entailed by inclusions

$$[x+a(t+1)(R(x)+G(t+1,x,\omega)) \notin C] \subset [|R(x)+G(t+1,x,\omega)| > \frac{d(x,\partial C)}{a(t+1)}] \subset$$
$$\subset [|G(t+1,x,\omega)| > \frac{d(x,\partial C)}{a(t+1)} - R_0].$$

Further, observe that

(8) $$|R^*(t,x)-R(x)|^2 = |E((R(x)+G(t+1,x,\omega))1_{[x^*\notin C]})|^2 \leqq$$
$$\leqq E|R(x)+G(1,x,\omega)|^2 P(x^*\notin C) \leqq KP(x^*\notin C).$$

Fix a β such that $0<\beta<d(\theta,\partial C)$ and that the Lipschitz continuity required in (ix) and (x) is fulfilled in $C\cap U_\beta(\partial C)$. Choose a $D>0$ so that

(9) $$\sup_{x\in C\cap U_\beta(\partial C)} P(|G(1,x,\omega)| > D-R_0) < \gamma^2/(4K),$$

where

$$\gamma = \sup_{x\in C\cap U_\beta(\partial C)} \langle R(x), x-\theta\rangle \,/\, \sup_{x\in\partial C}|x-\theta|, \qquad (\gamma<0);$$

let t_0 be such that

(10) $$R_0 a(t+1) \leqq \beta/2, \quad Da(t+1) < \beta, \quad \forall t \geqq t_0.$$

For $t\geqq t_0$ define a partition $\{C_1, C_{2t}, C_{3t}\}$ of C as follows:

$$C_1 = C\cap U_\beta^c(\partial C), \quad C_{2t} = C\cap U_\beta(\partial C)\cap U_{Da(t+1)}^c(\partial C),$$
$$C_{3t} = C\cap U_{Da(t+1)}(\partial C).$$

For $x \in C_1$ (and $t \geq t_0$) we have

$$P(x^* \notin C) \leq P(|G(1,x,\omega)| > \frac{\beta}{2a(t+1)}) \leq 4K\,\beta^{-2}\,a^2(t+1),$$

according to (7), (10) and Čebyšev inequality; inserting this into (8) we get

$$|R^*(t,x) - R(x)| \leq 2K\,\beta^{-1}a(t+1),$$

hence

$$(11)\quad |a(t+1)\langle R^*(t,x),x-\theta\rangle - a(t+1)\langle R(x),x-\theta\rangle| \leq K_1 a^2(t+1).$$

For $x \in C_{2t}$ we have, according to (9),

$$P(x^* \notin C) \leq P(|G(1,x,\omega)| > D-R_0) \leq \gamma^2/(4K),$$

hence

$$|R^*(t,x) - R(x)| \leq |\gamma|/2,$$

and, making use of the definition of γ,

$$(12)\qquad a(t+1)\langle R^*(t,x),x-\theta\rangle \leq (1/2)a(t+1)\langle R(x),x-\theta\rangle .$$

Finally, let $x \in C_{3t}$. Then there exist $x' \in \partial C$ and $0 < \delta < D$ such that

$$x = x' - \delta\, a(t+1)\dot{F}(x')/|\dot{F}(x')| .$$

Let us represent $G(1,x,\omega)$ as $B_x^{1/2}\,Y$, where Y has the standard $N(0,I)$ normal distribution. Taking ε from (xi), we have

$$P(|Y| \geq a^{-\varepsilon}(t+1)) \leq a^2(t+1), \quad \forall t \geq t_\varepsilon ;$$

assume $t_0 \geq t_\varepsilon$. We get

$$(13)\qquad R^*(t,x) = E((R(x) + B_x^{1/2}\,Y)\,1_{[x^*\in C]\cap[|Y|<a^{-\varepsilon}(t+1)]}) +$$

$$+ \text{ a term less than } K^{1/2}a(t+1) \text{ in the norm.}$$

The leading term in (13) can be expressed as

$$(14)\quad E((R(x')+B_{x'}^{1/2}\,Y + a(t+1)(\vartheta_1 + \vartheta_2\,Y)).$$

$$\cdot 1_{[x'+a(t+1)(R(x')+B_{x'}^{1/2}Y-\delta\,\dot{F}(x')/|\dot{F}(x')|)+a^2(t+1)(\vartheta_1+\vartheta_2 Y)\in C]}\cdot$$

$$\cdot 1_{[|Y|<a^{-\varepsilon}(t+1)]}),$$

where the vector ϑ_1 and the matrix ϑ_2 (though depending on x and t) are bounded by constants. The event $[\eta \in C]$ is the same as $[F(\eta) \leq 0]$; using Taylor expansion, we can express the event in the 2nd line of (14) as

$$(15)\quad [\langle \dot{F}(x'),\rho\rangle + (1/2)a(t+1)\langle \ddot{F}\rho,\rho\rangle \leq 0],$$

where ρ stands for

$$(16)\qquad R(x') + B_{x'}^{1/2} Y - \delta \dot{F}(x') / |\dot{F}(x')| + a(t+1)(\vartheta_1 + \vartheta_2 Y)$$

and $\ddot{F} = \ddot{F}(x' + \vartheta a(t+1)\rho)$, $0<\vartheta<1$. Replacing the terms with the coefficient $a(t+1)$ by zero in (16), (15) and in the 1st line of (14), and the indicator in the 3rd line by one, i.e. replacing $R^*(t,x)$ by

$$(17)\qquad E((R(x')+B_{x'}^{1/2} Y)\, 1_{[\langle \dot{F}(x'), R(x')+B_{x'}^{1/2}Y-\delta\dot{F}(x')/|\dot{F}(x')|\rangle \leqq 0]})$$

we change $a(t+1)\langle R^*(t,x), x-\theta\rangle$ at most by $K_2 a^{2-\varepsilon}(t+1)$, as follows by an easy calculation.

Let us continue investigating the expression (17); write it as

$$B_{x'}^{1/2} E((B_{x'}^{-1/2} R(x') + Y)\ .$$

$$\cdot 1_{[\langle B_{x'}^{1/2}\dot{F}(x'), B_{x'}^{-1/2}R(x')+Y-\delta B_{x'}^{-1/2}\dot{F}(x')/|\dot{F}(x')|\rangle \leqq 0]}) =$$

$$= B_{x'}^{1/2}\ E(Z 1_{[\langle B_{x'}^{1/2}\dot{F}(x'), Z\rangle \leqq \delta|\dot{F}(x')|]}),$$

where Z has the $N(B_{x'}^{-1/2}R(x'), I)$ normal distribution. The latter expression further equals

$$(18)\qquad B_{x'}^{1/2}(B_{x'}^{-1/2}R(x') - \varkappa_1 B_{x'}^{1/2}\dot{F}(x'))\,\varkappa_2 = (R(x') - \varkappa_1 B_{x'}\dot{F}(x'))\,\varkappa_2$$

for some $\varkappa_1 > 0$, $0<\varkappa_2<1$, because of the following lemma (which is easy to verify):

<u>Lemma</u>. Let Z have the $N(a,I)$ normal distribution; then

$$E(Z 1_{[\langle b,Z\rangle \leqq c]}) = \left(a - \frac{\varphi(\alpha)\, b}{\Phi(\alpha)|b|}\right)\Phi(\alpha),$$

where $\alpha = (c - \langle a,b\rangle)/|b|$, $a,b \in \mathbb{R}^n$, $c \in \mathbb{R}$, φ and Φ the $N(0,1)$ density and distribution function.

The $\varkappa_1$'s in (18) depend on x' and δ ,i.e. on x and t; however, it folows from the lemma that

$$\varkappa_2(x,t) \geqq \varkappa_{20} > 0, \qquad \forall x \in C_{3t},\quad t \geqq t_o.$$

Thus we have (still for $x \in C_{3t}$, $t \geqq t_o$)

$$(19)\qquad a(t+1)\langle R^*(t,x), x-\theta\rangle \leqq \varkappa_{20}\, a(t+1)\langle R(x'), x'-\theta\rangle + K_3 a^{2-\varepsilon}(t+1) \leqq$$

$$\leqq \varkappa_{20} a(t+1)\langle R(x), x-\theta\rangle + K_4 a^{2-\varepsilon}(t+1).$$

Combining results for C_1, C_{2t}, C_{3t}, i.e. (11),(12),(19), we get

$$a(t+1)\ \langle R^*(t,x), x-\theta\rangle \leq \min(1/2, \varkappa_{20})a(t+1)\ \langle R(x), x-\theta\rangle +$$
$$+ K_5 a^{2-\varepsilon}(t+1), \quad \forall x \in C, \quad t \geq t_0.$$

Inserting this into (6), we can see that the same bound is valid for LV(t,x). Hence, the assertion of Theorem 1 follows (see Nevel'son, Has'minskiĭ (1972), Theor.2.7.1).

In comparison with the situation of Theorem 1, the redefinition (5) of $X^*(t)$ means only a replacement of $G(1,x,\omega)$ by $G(1,x,\omega)+ + H(1,\omega)$ (in a neighbourhood of ∂C) and that of B_x by B_x+bI. The condition (xiii) ensures then that (xii) is fulfilled for the new covariance matrix B_x+bI. Thus the assertion of Theorem 2 follows from Theorem 1.

References

Kushner,H.J., Clark,D.S. (1978). Stochastic Approximation Methods for Constrained and Unconstrained Systems. Springer Verlag,New York.

Nevel'son,M.B., Has'minskiĭ,R.Z. (1972). Stohastičeskaja Approksimacija i Rekurentnoe Ocenivanie. Nauka, Moscow.
English translation: Stochastic Approximation and Recursive Estimation, Translations of Mathem.Monographs, Vol.47,American Mathem. Soc.,Providence (1976).

LEARNING AUTOMATON FOR FINITE SEMI-MARKOV DECISION PROCESSES

Yousri M. El-Fattah
L.E.E.S.A.
Faculté des Sciences
B.P. 1014, Rabat
Morocco

A finite semi-Markov decision process is studied to maximize the expected average reward. The semi-Markov kernel of the process depends on an unknown parameter taking values in a subset $[\underline{a}, \underline{b}]$ of $\mathbb{R}^s$. A controller modelled as a learning automaton updates sequentially the probabilities of generating decisions based on the observed decisions, states, and jump times. Convergence results are stated in the form of theorems and some examples are given.

I. Adaptive Control Problem

Let I denote the set of states of a controlled system. The elements of I are numbered 1 through N. The system's state is observed in continuous time $t \geq 0$. For each state $i \in I$ there exists a finite number M_i of possible actions. Let the action k be taken according to a system of probability vectors $D = \{\underline{d}^i\}$, $i \in I$, where $\underline{d}^i$ is a row M_i vector whose k-th component d_k^i specifies the probability of taking the action k whenever the system is in state i. Henceforth, D is called a control policy. Note that $\underline{d}^i$ must belong to the simplex

$$S^{M_i} = \{\underline{d}^i \in R^{M_i},\ d_k^i \geq 0,\ \sum_{k=1}^{M_i} d_k^i = 1\}. \tag{1}$$

For a control policy D the law of motion of the controlled system's state is given by the system of probabilities

$$Q(i,j,t,\underline{d}^i) \geq 0, \quad \sum_{j \in I} Q(i,j,\infty,\underline{d}^i) = 1, \text{ all } i \in I,\ \underline{d}^i \in S^{M_i}. \tag{2}$$

$Q(i,j,t,\underline{d}^i)$ denotes the probability that after making a transition into state i the system's state next makes a transition into state j in an amount of time less than or equal to t.

Let J_o denote the initial state of the system and for $n \geq 1$, let J_n denote the state of the system immediately after the nth transition has occurred. It is clear that for an arbitrary policy D the process $\{J_n, n=0,1,2,\ldots\}$ is a homogeneous Markov chain with transition probabilities

$$P(i,j,\underline{d}^i) = Q(i,j,\infty,\underline{d}^i) \tag{3}$$

If $P(i,j,\underline{d}^i) = 0$ for some pair (i,j) then $Q(i,j,t,\underline{d}^i) = 0$ for all $t > 0$ and we define the ratio $Q(i,j,t,\underline{d}^i)/P(i,j,\underline{d}^i)$ to be unity. With this convention we define

$$F(i,j,t,\underline{d}^i) = Q(i,j,t,\underline{d}^i)/\ P(i,j,\underline{d}^i) \tag{4}$$

Then for each pair (i,j) the function $F(i,j,.,\underline{d}^i)$ is a distribution function representing the conditional probability that a transition will take place within an amount of time t , given that the system's state has just entered i and will next enter j. The expected amount of time spent in a state i during each visit, given that the next state entered is j, is

$$\bar{t}(i,j,\underline{d}^i) = \int_0^\infty t\ F(i,j,dt,\underline{d}^i) \tag{5}$$

The expected amount of time spent in i during each visit is given by

$$\nu(i,\underline{d}^i) = \sum_{j\in I} \bar{t}(i,j,\underline{d}^i)\ P(i,j,\underline{d}^i). \tag{6}$$

We assume a certain reward structure is superimposed on the choice of control decisions. Whenever the system is in state i and action k is taken, an immediate reward R(i,k) is earned and, in addition, a reward rate r(i,k) is accumulated until the next transition occurs. That is, if a transition occurs after t units, then the total reward is given by R(i,k)+t r(i,k). For any stationary policy D the expected earning per transition for state i is

$$q(i,\underline{d}^i) = \sum_{k=1}^{M_i} (R(i,k)+\bar{t}(i,\delta_k)r(i,k))\ d^i_k \tag{7}$$

where δ_k denotes the Dirac measure concentrated at point k.

At this point let us introduce the following assumptions:

A.1. For all policies D the embedded Markov chain $\{J_n\}$, n=0,1,2,.. with the transition probabilities (3),has a single ergodic class consisting of all the states.

A.2. For all policies D the mean sojourn times (6) for all states i∈I are finite.

Based on A.1 there exists an invariant probability measure $\underline{\pi}$ on I , given by the equation

$$\underline{\pi}\ B = \underline{e}_1 \tag{8}$$

where $\underline{\pi} = (\ \pi_1\ \pi_2\ \dots\ \pi_N)$, B is the NxN matrix

$$B = (B_{ij})$$

$$B_{ij} \begin{cases} = 1 & j=1 & \text{all } i \\ = 1 - P(i,i,\underline{d}^i), & j=i, \quad j=2,..,N \\ = - P(i,j,\underline{d}^i), & j\neq i, \quad j=2,..,N & \text{all } i, \end{cases} \tag{9}$$

and $\underline{e}_1$ is the N row vector (1 0 ... 0). B is invertible due to the fact that it is of rank N. Let G be the inverse of B. It follows from (8) that

$$\underline{\pi} = \underline{e}_1\ G. \tag{10}$$

That is, $\underline{\pi}$ is the top row of G.

Under A.1, A.2,it is possible to show that the expected average reward (the long-run expected reward per unit time) can be expressed as

$$g(D) = g_1(D)/g_2(D) \tag{11}$$

where

$$g_1(D) = \sum_{i\in I} \pi_i \sum_{k=1}^{M_i} q(i,\delta_k)\, d_k^i$$

$$g_2(D) = \sum_{i\in I} \pi_i \sum_{k=1}^{M_i} (i,\delta_k)\, d_k^i \,.$$

It is assumed that the semi-Markov kernels (2) be specified up to a constant but unknown parameter $\underline{\theta}\in[\underline{a},\ \underline{b}]\subset \mathbb{R}^s$, so that

$$Q(i,j,t,\underline{d}^i) := Q(\underline{\theta}\ ;i,j,t,\underline{d}^i) \tag{12}$$

all $i,j\in I$, all policies D. The adaptive control problem can be stated as follows: It is required to determine strategies for estimating the parameter $\underline{\theta}$ and experimenting with the control policies D, based on the entire history of the system up to the current jump time, so as to approach asymptotically the maximum expected average reward.

II. Conditions Of Optimality

Let $\alpha\in I$ be arbitrary and $\beta\in\{1,..,M_\alpha\}$. Define the system of N row vectors $\underline{b}_{\alpha\beta}$ given by

$$\underline{b}^j_{\alpha\beta} \begin{cases} =0, & j=1 \\ =-\dfrac{M_\alpha}{M_\alpha-1}(P(\alpha,j,\delta_\beta)-\bar{P}(\alpha,j)), & j=2,..,N \end{cases} \tag{13}$$

where $\bar{P}(\alpha,j)$ denotes the arithmetic mean of the quantities $\{P(\alpha,j,\delta_\beta)\}$. Let $\underline{v}_{\alpha\beta}$ be the N row vector satisfying the equation

$$\underline{v}_{\alpha\beta}\, B = \underline{b}_{\alpha\beta} \,. \tag{14}$$

Using the variational approach ,see El-Fattah (1981), it is possible to prove the following Taylor-like expansion formula:

$$g(D') - g(D) = \sum_{\alpha,\beta} \frac{M_\alpha-1}{M_\alpha} \frac{\partial\ g}{\partial\ d^\alpha_\beta}(D)\ (d^{\alpha\prime}_\beta - d^\alpha_\beta) + O(\sum_{\alpha,\beta} |d^{\alpha\prime}_\beta - d^\alpha_\beta|^2), \tag{15}$$

where

$$\partial g/\partial\, d^\alpha_\beta = (g_2 \frac{\partial\ g_1}{\partial\ d^\alpha_\beta} - g_1 \frac{\partial\ g_2}{\partial\ d^\alpha_\beta})/\ (g_2)^2 \tag{16a}$$

$$\partial g_1/\partial d^\alpha_\beta = (\frac{M_\alpha}{M_\alpha-1}(q(\alpha,\delta_\beta) - \bar{q}(\alpha)) - \underline{v}_{\alpha\beta}\ \underline{\eta}\ ) \tag{16b}$$

$$\partial g_2/\partial d^\alpha_\beta = (\frac{M_\alpha}{M_\alpha-1}(\nu(\alpha,\delta_\beta) - \bar{\nu}(\alpha)) - \underline{v}_{\alpha\beta}\ \underline{\kappa}\ ) \tag{16c}$$

$\bar{q}(\alpha)$, $\bar{\nu}(\alpha)$ denote the arithmetic means of the quantities $\{q(\alpha,\delta_\beta)\}$, $\{\nu(\alpha,\delta_\beta)\}$, respectively. $\underline{\eta}$ and $\underline{\kappa}$ denote the column N vectors with the respective components:

$$\eta^i = \sum_{k=1}^{M_i} d_k^i\, q(i,\delta_k)\ , \qquad \kappa^i = \sum_{k=1}^{M_i} d_k^i\, \nu(i,\delta_k). \tag{17}$$

Let $\delta_{\alpha\beta}(\Delta)$ designate the following variation of a policy D:

$$D+\delta_{\alpha\beta}(\Delta) := \begin{cases} \text{Change the probability vector} \\ \underline{d}^\alpha \text{ to } \underline{d}^\alpha + \Delta\, \underline{z}_{\alpha\beta}\text{; let the remaining} \\ \text{probabilities be the same.} \end{cases} \tag{18}$$

where

$$\underline{z}_{\alpha\beta} = (\underbrace{-\frac{1}{M_\alpha-1}, \ldots, -\frac{1}{M_\alpha-1}}_{\beta-1}, 1, -\frac{1}{M_\alpha-1}, \ldots, -\frac{1}{M_\alpha-1}) \tag{19}$$

Naturally Δ must be admissible, i.e., $\underline{d}^\alpha + \Delta \underline{z}_{\alpha\beta}$ must remain in S^{M_α}.

Theorem 1. Let D be an arbitrary policy and $\delta_{\alpha\beta}(\Delta)$ be an admissible variation (18) such as

$$\Delta = \left(g_2 \frac{\partial g_1}{\partial d_\beta^\alpha} - g_1 \frac{\partial g_2}{\partial d_\beta^\alpha} \right), \quad |\Delta| > 0\ , \tag{20}$$

and

$$0 < \gamma < \bar{\gamma}(\alpha, D). \tag{21}$$

Then the resulting variation of the expected average reward is strictly positive,

$$\delta_{\alpha\beta}\, g(D) > 0. \tag{22}$$

Theorem 2. If D^o is an optimal policy, then it is necessary that

$$(d_\beta^\alpha - d_\beta^{\alpha o})\left(g_2(D^o) \frac{\partial g_1}{\partial d_\beta^\alpha}(D^o) - g_1(D^o) \frac{\partial g_2}{\partial d_\beta^\alpha}(D^o) \right) \leqslant 0 \quad \text{all } \alpha,\beta\ . \tag{23}$$

Theorem 3. Let there exist a deterministic policy D^o,

$$d^{\alpha o} = \delta_{\beta_\alpha} \quad \text{all } \alpha \in I \tag{24}$$

such that

$$\max_\beta \left(g_2(D^o) \frac{\partial g_1}{\partial d_\beta^\alpha}(D^o) - g_1(D^o) \frac{\partial g_2}{\partial d_\beta^\alpha}(D^o)\right) = g_2(D^o) \frac{\partial g_1}{\partial d_{\beta_\alpha}^\alpha}(D^o) - g_1(D^o) \frac{\partial g_2}{\partial d_{\beta_\alpha}^\alpha}(D^o), \text{ all } \alpha \in I. \tag{25}$$

Then, D^o is optimal.

Example 1. This example is due to Jewell (1963). Consider the two state problem of a machine that is running (state 1) or has broken down (state 2). If the machine is running there are two maintenance alternatives:

Alternative 1: $R(1,1)=0$, $r(1,1)=$ \$100/ day, $\bar{t}(1,2,\delta_1)=4$ days, $P(1,2,\delta_1)=1$;

Alternative 2: $R(1,2)=0$, $r(1,2)=$ \$84/day, $\bar{t}(1,2,\delta_2)=5$ days, $P(1,2,\delta_2)=1$.

If the machine has broken down, there are two repair alternatives:
Alternative 1: $R(2,1)=0$, $r(2,1)=-\$65$/day, $\bar{t}(2,1,\delta_1)=4$days, $P(2,1,\delta_1)=1$;
Alternative 2: $R(2,2)=-\ 100$, $r(2,2)=-\$200$/day, $\bar{t}(2,1,\delta_2)=1$day, $P(2,1,\delta_2)=1$.

Alternative 2 may be thought of as an outside repairman whose expensive fixed and running charges are compensated for by his quick service time. It follows from (6),(7) that

$q(i,\delta_k)$

i \ k	1	2
1	400	420
2	-260	-300

$\nu(i,\delta_k)$

i \ k	1	2
1	4	5
2	4	1

Let the superscript AB (A,B=1,2) denote the deterministic policy of taking actions A and B in states 1 and 2 respectively. By direct evaluation using (11) we obtain

$$g^{11}= 17.5, \quad g^{12}=20, \quad g^{21}=17.77, \quad g^{22}=20$$

in dollars per day so that either (A,B) or (B,B) are the maximal gain rate stationary policies, indicating that expensive repair should always be used!

For any policy D,the gradients (16) can be tabulated as follows:

$\partial g_1/\partial d^{\alpha}_{\beta}$

α \ β	1	2
1	-10	10
2	20	-20

$\partial g_2/\partial d^{\alpha}_{\beta}$

α \ β	1	2
1	-0.5	0.5
2	1.5	-1.5

For the determinstic policy BB we have $g_1=60$, $g_2=3$ and

$$g_2 \frac{\partial g_1}{\partial d^{\alpha}_{\beta}} - g_1 \frac{\partial g_2}{\partial d^{\alpha}_{\beta}}$$

tabulated as follows:

α \ β	1	2
1	0	0
2	-30	30

Hence, following Theorem 3, the policy BB is optimal.

Example 2. (The Streetwalker's Dilemma)

The following is taken from Ross (1970). Consider a prostitute working in the city of Paris, and suppose that potential customers arrive in accordance with a Poisson process with rate λ. Each potential customer makes an offer consisting of the pair (i,F_i),where i is the amount of money offered and F_i is the distribution of the amount of time that the prostitute must spend with the customer. If the offer is rejected, then the arrival leaves and the prostitute waits for the next potential customer. If the offer is accepted, then all potential customers who arrive while the prostitute is busy are assumed lost. The successive offers are assumed independent and the offer (i,F_i) occurs with probability P_i, where $\sum_{i=1}^{N} P_i = 1$. The prostitute's dilemma is to choose her customers so as to maximize the long-run average return.

The above may be viewed as a two-action semi-Markov decision process with states 1,2,..,N where state i means that the prostitute must decide whether or not to accept an offer of (i,F_i). If we let $\bar{t}_i = {}_0\int^{\infty} x dF_i(x)$ be the mean time spent with an i-type customer, and let action 1 be the "accept" and action 2 the "reject" action, then the parameters of the process are given by

$$P(i,j,\delta_1) = P_j \qquad\qquad P(i,j,\delta_2) = P_j$$

$$\nu(i,\delta_1) = \bar{t}_i + \frac{1}{\lambda} \qquad\qquad \nu(i,\delta_2) = \frac{1}{\lambda}$$

$$q(i,\delta_1) = i \qquad\qquad q(i,\delta_2) = 0$$

Hence,

$$g_1(D) = \sum_{i=1}^{N} P_i\, d_1^i\, i$$

$$g_2(D) = \frac{1}{\lambda} + \sum_{i=1}^{N} P_i\, d_1^i\, \bar{t}_i$$

$$\partial g_1/\partial d_\beta^\alpha \begin{cases} = \alpha P_\alpha, & \beta=1 \\ = -\alpha P_\alpha, & \beta=2 \end{cases}, \qquad \partial g_2/\partial d_\beta^\alpha \begin{cases} = \bar{t}_\alpha P_\alpha, & \beta=1 \\ = -\bar{t}_\alpha P_\alpha, & \beta=2 \end{cases}.$$

According to Theorem 3, the deterministic policy (24) is optimal if

$$\left(\frac{1}{\lambda} + \sum_{i:\beta_i=1} P_i \bar{t}_i\right)\alpha P_\alpha - \left(\sum_{i:\beta_i=1} P_i\, i\right)\bar{t}_\alpha P_\alpha \begin{cases} > 0 & \beta=1 \\ < 0 & \beta=2 \end{cases}.$$

Therefore, the rule : Accept an offer if

$$\frac{\alpha}{\bar{t}_\alpha} > \frac{\sum_{i:\beta_i=1} i P_i}{\frac{1}{\lambda} + \sum_{i:\beta_i=1} P_i \bar{t}_i} = g(D^o) \quad \text{all } \alpha \in I.$$

III. Recursive Parameter Estimation

Suppose that the system of semi-Markov kernels (12) admits the representation

$$Q(\underline{\theta};\, i,j,t,\delta_k) = {}_0\int^{t} q(\underline{\theta},\, i,k,j,\tau)\, d\tau \tag{26}$$

It is assumed that there exists $\varepsilon>0$ such that for every i,j I either $q(\underline{\theta},i,k,j,\tau)>\varepsilon$ for all k, $\underline{\theta}$, or $q(\underline{\theta},i,k,j,\tau)=0$ for all k, $\underline{\theta}$.

From now on it will be considered that the sequence of control actions A_n, n=0, 1,2,.., is determined sequentially as follows: J_o, $\underline{\theta}_o$, and D_o are arbitrary. Let the present epoch be n, i.e., the system has just affected n transitions, n=0,1,2,.. A_n is the random variable generated with the probability $P(A_n=k) = d_k^{J_n}(n)$, $k\in\{1,..,M_{J_n}\}$

Applying the control A_n, the system's state will transit from J_n to J_{n+1}. After a random interval T_{n+1} $\underline{\theta}_n$ is then updated to $\underline{\theta}_{n+1}$, as below. Knowing $\underline{\theta}_{n+1}$, J_n, A_n and D_n, the control policy will also be updated at the next epoch to D_{n+1}. If F_n denotes the minimum σ-algebra generated by J_o, T_o, J_1, T_1,..., J_{n-1}, T_{n-1}, J_n then D_n is a random variable measurable in F_n. Only randomized policies, $D_n \in S_\varepsilon$, are allowed where

$$S_\varepsilon = \{D: \underline{d}^\alpha \; S^M_\varepsilon\alpha \quad \text{all } \alpha \}$$
$$S^M_\varepsilon\alpha = \{ \underline{d}^\alpha: \; \underline{d}^\alpha \in S^M\alpha \quad \text{and} \quad d^\alpha_\beta \geq \varepsilon \quad \text{all } \beta \} \tag{27}$$

We propose the following recursive maximum likelihood estimation procedure:
If $J_n = i$, $A_n = k$, $J_{n+1} = j$, $T_{n+1} = t_{n+1}$, then

$$\underline{\theta}_{n+1} = \underline{\theta}_n + \frac{\gamma}{n_k+1} \, \frac{\nabla q(\underline{\theta}_n, i,k,j,t_{n+1})}{q(\underline{\theta}_n, i,k,j,t_{n+1})}, \quad n=0,1,2,.. \qquad \gamma > 0, \tag{28}$$

where n_k denotes the number of epochs the decision k is taken up to n-1.

Let us now investigate the convergence properties of (28). For that purpose we shall make use of the following assumptions:

A.3. For each state-control pair (i,k) there corresponds a subset $K_{ik} \subseteq K=\{1,..,s\}$ which may be void, such that $\bigcup_{i,k} K_{ik} = K$, and the conditions

$$\sum_{\ell \in K_{ik}} G_\ell(i,k,\underline{\theta})(\theta^\ell - \theta^{\ell *}) \leq - \rho(i,k,\underline{\theta}) \sum_{\ell \in K_{ik}} (\theta^\ell - \theta^{\ell *})^2$$
$$\sum_{\ell \in K/K_{ik}} G_\ell(i,k,\underline{\theta})(\theta^\ell - \theta^{\ell *}) = 0 \tag{29}$$

hold, where

$$\underline{G}(i,k,\underline{\theta}) = E\left(\frac{\nabla q(\underline{\theta}, J_n, k, J_{n+1}, \tau_{n+1})}{q(\underline{\theta}, J_n, k, J_{n+1}, \tau_{n+1})} \Big/ J_n = i \right)$$
$$= \sum_{j \in I} \int_0^\infty \frac{\nabla q(\underline{\theta}, i, k, j, t)}{q(\underline{\theta}, i, k, j, t)} \; q(\underline{\theta}^*, i,k,j,t)\, dt \tag{30}$$

and $\rho(i,k,\underline{\theta})$ is a continuous function on $[\underline{a}, \underline{b}]$, all i,k. $\underline{\theta}^*$ denotes the true parameter in $[\underline{a}, \underline{b}]$.

Noting that D_n is F_n-measurable, it follows from (30) that

$$E\left(\frac{\nabla q(\underline{\theta}_n, J_n, A_n, J_{n+1}, T_{n+1})}{q(\underline{\theta}_n, J_n, A_n, J_{n+1}, T_{n+1})} \Big/ F_n \right) = \sum_{k=1}^{M_i} \underline{G}(i,k,\underline{\theta}_n)\, d^i_k(n)\; 1(J_n = i).$$

Hence A.3 implies that for any state-action pair (i,k), whenever $J_n = i$ and $A_n = k$, the "forcing term" $\nabla q / q$ in the learning algorithm (28) is a pseudo-gradient in the sense of Polyak and Tsypkin (1973) with respect to the function $V(\underline{\theta}) = - \| \underline{\theta} - \underline{\theta}^* \|^2$, i.e.,

$$\nabla^T V(\underline{\theta}_n)\, E\left(\frac{\nabla q(\underline{\theta}_n, J_n, A_n, J_{n+1}, T_{n+1})}{q(\underline{\theta}_n, J_n, A_n, J_{n+1}, T_{n+1})} \Big/ F_n \right) \geq 0 .$$

A.4. For each (i,k) there exist the positive continuous functions $\phi(i,k,\underline{\theta})$, $\psi(i,k,\underline{\theta})$ such that

$$\text{trace } W(i,k,\underline{\theta}) \leq \phi(i,k,\underline{\theta}) + \psi(i,k,\underline{\theta}) \sum_{\ell \in K_{ik}} (\theta^{\ell} - \theta^{\ell *})^2 \tag{31}$$

where W denotes the matrix

$$W(i,k,\underline{\theta}) = E\left(\frac{\nabla q(\underline{\theta}, J_n, k, J_{n+1}, T_{n+1})\, \nabla^T q(\underline{\theta}, J_n, A_n, J_{n+1}, T_{n+1})}{(q(\underline{\theta}, J_n, k, J_{n+1}, T_{n+1}))^2} \Big/ J_n = i \right)$$

$$= \sum_{j \in I} \int_0^{\infty} \frac{\nabla q(\underline{\theta}, i,k,j,t)\, \nabla^T q(\underline{\theta}, i,k,j,t)}{(q(\underline{\theta}, i,k,j,t))^2}\, q(\underline{\theta}^*, i,k,j,t)\, dt \tag{32}$$

all $i \in I$, $k \in \{1,..,M_i\}$.

Example 3. Consider the controlled Markov jump process with the kernel

$$Q(\underline{\theta}, i,j,t,\delta_k) = P(\underline{\theta}, i,k,j)(1 - \exp(-\lambda_i(\underline{\theta},k)t))$$

$$P(\underline{\theta}, i,k,i) = 0 \quad \text{all } i,k.$$

It is obvious that

$$q(\underline{\theta}, i,k,j,t) = \lambda_i(\underline{\theta},k)\, P(\underline{\theta}, i,k,j)\, \exp(-\lambda_i(\underline{\theta},k)).$$

The recursive estimation algorithm (28) becomes

If $J_n = i$, $A_n = k$, $J_{n+1} = j$, $T_{n+1} = t_{n+1}$ then

$$\underline{\theta}_{n+1} = \underline{\theta}_n + \frac{\gamma}{n_k + 1} \left(\frac{\nabla P(\underline{\theta}_n, i,k,j)}{P(\underline{\theta}_n, i,k,j)} + \left(\frac{1}{\lambda_i(\underline{\theta}_n,k)} - t_{n+1} \right) \nabla \lambda_i(\underline{\theta}_n,k) \right), \quad n=0,1,2,..$$

It is easy to see that

$$G(i,k,\underline{\theta}) = \sum_{j \in I} \frac{\nabla P(\underline{\theta}, i,k,j)}{P(\underline{\theta}, i,k,j)} P(\underline{\theta}^*, i,k,j) + \nabla \lambda_i(\underline{\theta},k) \left(\frac{1}{\lambda_i(\underline{\theta},k)} - \frac{1}{\lambda_i(\underline{\theta}^*,k)} \right).$$

It is assumed that

$$0 < c_1 \leq \lambda_i(\underline{\theta},k) \leq c_2 < \infty$$

uniformly for all i,k, $\underline{\theta}$.

For simplicity, consider the uncontrolled case ($M_i = 1$) and let $I = \{1,2\}$. Let $\lambda_i(\underline{\theta}) = \theta^i$. Then

$$G(1,\underline{\theta}) = ((\theta^1 - \theta^{1*}) / \theta^1 \theta^{1*},\ 0),\quad G(2,\underline{\theta}) = (0,\ (\theta^2 - \theta^{2*}) / \theta^2 \theta^{2*}).$$

Hence, A.3 is satisfied with $K_1 = \{1\}$, $K_2 = \{2\}$, and

$$\rho(1,\underline{\theta}) = 1/\theta^1 \theta^{1*}, \quad \rho(2,\underline{\theta}) = 1/\theta^2 \theta^{2*}.$$

A.4 is also satisfied.

Theorem 4. If assumptions A1-A4 hold, then for any arbitrary randomized policies $\{D_n\}$, $n = 0,1,2,..$ such that D_n is F_n-measurable and $D_n \in S_\varepsilon$, the estimate $\underline{\theta}_n$ converges to $\underline{\theta}^*$ almost surely, and

$$\sum_{n=0}^{\infty} \frac{1}{n+1} \sum_{i\in I} \sum_{\ell\in K_i} (\theta_n^{\ell} - \theta^{\ell *})^2 \, 1(J_n = i) < \infty \qquad \text{almost surely} \tag{33}$$

where $1(.)$ is the indicator function of (.), and $K_i = \bigcup_{k=1}^{M_i} K_{ik}$.

IV. Recursive Adaptive Control

Let us consider a control decision model in the form of a stochastic automaton which experiments with control policies while observing the process state transitions at consecutive epochs. The automaton starts with an arbitrary random policy D_0, e.g., the policy generating equally probable decisions for each observed state, i.e., $d_k^i(0) = 1/M_i$ for all i,k. Let the present epoch be n (this denotes that the nth transition has taken place). Let the current process state be $J_n = \alpha$. According to the current policy D_n, the automaton generates a control decision β with probability $d_\beta^\alpha(n)$, $\beta \in \{1,..,M_\alpha\}$. The chain then makes a transition to a j-th state after a random interval T_{n+1}. The parameter estimation is updated following the scheme (28). Then the policy adaptation device, using the updated estimate of $\underline{\theta}$, calculates the gradient $\partial g/\partial d_\beta^\alpha$ corresponding to the estimate $\underline{\theta}_{n+1}$ and updates the control policy according to the following reinforcement scheme:

If $J_n = \alpha$, $A_n = \beta$, then

$$D_{n+1} = D_n + \delta_{\alpha\beta}(\bar{\Delta}_{\alpha\beta}(n)), \quad n=0,1,2,...$$

where $\bar{\Delta}_{\alpha\beta}(n)$ is some function of the gradient $\partial \breve{g}/\partial d_\beta^\alpha$ that preserves sign. Here the superscript $\smile$ denotes that the true parameter $\underline{\theta}^*$ is replaced by its current estimate $\underline{\theta}_{n+1}$ in calculating the gradient

$$\frac{\partial \breve{g}}{\partial d_\beta^\alpha}(D_n) = \frac{\partial g}{\partial d_\beta^\alpha}(D_n, \underline{\theta}_{n+1}).$$

Moreover, $\bar{\Delta}$ is subject to the constraint that $\underline{d}^\alpha(n+1) \in S_\varepsilon^{M_\alpha}$ (Eqn.(27)). That condition is satisfied when

$$-\Delta_{mx}^- \leqslant \bar{\Delta}_{\alpha\beta} \leqslant \Delta_{mx}^+$$

where

$$\Delta_{mx}^+(\varepsilon) = \min\left(1 - d_\beta^\alpha - (M_\alpha - 1)\varepsilon\,,\; (M_\alpha - 1)\min_{k\neq\beta}(d_k^\alpha - \varepsilon)\right)$$

$$\Delta_{mx}^-(\varepsilon) = \min\left(d_\beta^\alpha - \varepsilon,\; (M_\alpha - 1)\left(1 - (M_\alpha - 1)\varepsilon - \max_{k\neq\beta} d_k^\alpha\right)\right) \tag{34}$$

A.5. For all α,β , the gradient $\partial \widetilde{g}/\partial d^{\alpha}_{\beta}$ can be represented as

$$\partial \widetilde{g}/\partial d^{\alpha}_{\beta} = F_{\alpha}(D,\underline{\theta})\ f_{\alpha\beta}(D,\underline{\theta}) \tag{35}$$

so that $F_{\alpha}(D,\underline{\theta})$ is a positive continuous function on $S_{\varepsilon}x[\underline{a},\underline{b}]$ and $f_{\alpha\beta}(D,\underline{\theta})$ satisfies the condition that there exists a constant c_1, so that the condition

$$\sum_{\beta=1}^{M_{\alpha}} (f_{\alpha\beta}(D,\underline{\theta}) - f_{\alpha\beta}(D,\underline{\theta}^*))^2 \leq c_1 \sum_{\ell\in K_{\alpha}} (\theta^{\ell}-\theta^{\ell *})^2 \tag{36}$$

uniformly on policies.

The following reinforcement scheme for the automaton controller is suggested: If $J_n=\alpha$, $A_n=\beta$, then

$$\underline{d}^i(n+1) \begin{cases} = \underline{d}^i(n), & \text{if } i\neq \alpha \\ = \underline{d}^{\alpha}(n) + \bar{\Delta}_{\alpha\beta}\ \underline{z}_{\alpha\beta} & \text{if } i= \alpha \end{cases} \tag{37}$$

where

$$\bar{\Delta}_{\alpha\beta} \begin{cases} = \Delta^+_{mx} & \text{if } \Delta_{\alpha\beta} > \Delta^+_{mx} \\ = \Delta_{\alpha\beta} & \text{if } - \Delta^-_{mx} \leq \Delta_{\alpha\beta} \leq \Delta^+_{mx} \\ =- \Delta^-_{mx} & \text{if } \Delta_{\alpha\beta} <- \Delta^-_{mx} \end{cases} \tag{38}$$

and

$$\Delta_{\alpha\beta} = \frac{a}{n+1}\ f_{\alpha\beta}(D_n,\underline{\theta}_{n+1})\ ,\quad a> 0 \tag{39}$$

It is also possible to write (38) as

$$\bar{\Delta}_{\alpha\beta} = \frac{a}{n+1}\ \xi_{\alpha\beta}(\varepsilon)\ f_{\alpha\beta}(D_n,\underline{\theta}_{n+1}) \tag{40}$$

Note that $\xi_{\alpha\beta}=1$ if $\underline{d}^{\alpha}+ \bar{\Delta}_{\alpha\beta}\underline{z}_{\alpha\beta}$ still lies on the simplex $S^{M_{\alpha}}_{\varepsilon}$ $=0$ when $f_{\alpha\beta}>0$ and $\Delta^+_{mx}=0$, or $f_{\alpha\beta}<0$ and $\Delta^-_{mx}=0$.

Theorem 5. Let g^* be the expected average reward corresponding to the optimal policy D^o. Then, under A.1-A.5 the reinforcement scheme (37) through (39) for the learning automaton controller yields

1. $(g^* - g_n)^2$ converges almost surely to a random variable.
2. $\liminf_{n\to\infty} (g^* - g_n) \sum_{\beta=1}^{M_{\alpha}} \xi_{\alpha\beta}\ F_{\alpha}(D_n,\underline{\theta}^*)(f_{\alpha\beta}(D_n,\underline{\theta}^*))^2 = 0$ almost surely.

References

El-Fattah ,Y.M. (1981) Gradient approach for recursive estimation and control in finite Markon chains. Adv. Appl. Probability, 13, 778-803.

Jewell, W.S. (1963) Markov-renewal programming, I,II. Operations Research, 2 , 938-971.

Polyak, B.T. and Tsypkin, Ya.Z. (1973) Pseudo-gradient adptation and training algorithms. Automation and Remote Control, 34, 377-397.

Ross, S.M. (1970) Applied probability models with optimization applications. Holden Day, San Francisco.

A LOCAL ASYMPTOTIC MINIMAX OPTIMALITY OF AN ADAPTIVE ROBBINS MONRO STOCHASTIC APPROXIMATION PROCEDURE[1)]

Václav Fabian
Michigan State University
East Lansing, MI 48823

Results, obtained earlier, on asymptotic properties of an adaptive RM stochastic approximation method, are strengthened. The method is shown optimal in the sense of local asymptotic minimax risk and robust with respect to small changes of the unknown density and small changes of the estimated parameter.

1. Introduction

1.1. Summary of the result. The RM stochastic approximation method was proposed by Robbins and Monro (1951) for estimating a zero point θ of a function f, satisfying certain conditions, on the basis of unbiased estimates of the values of f at points selected by the approximation method. Considerable research followed. Abdelhamid (1973) and Anbar (1973) proposed a modification of the RM method for the case when the observation errors are distributed according to a known density g and showed that the method is asymptotically optimal in a classical sense. Fabian (1973) (F hereafter) showed that the same can be obtained if g is symmetric but unknown.

The classical optimality is a weak and dubious optimality. We shall strengthen the asymptotic result established in F for the adaptive RM method by showing that the convergence in distribution holds uniformly in shrinking neighborhoods of the parameter estimated and of the unknown density (part (i) of Theorem 3.3). We establish a local asymptotic minimax adaptive (LAMA) property of the adaptive RM method in part (ii) of Theorem 3.3 and infer limited optimality for the method in Remark 3.4.

1.2. Remark. The results and concepts obtained here are based on those in the asymptotic theory of estimation, due to Le Cam, Hájek and others, for locally asymptotically normal (LAN) problems. Fabian and Hannan (1982) (FH hereafter) proposed a definition of the locally asymptotically minimax (LAM) and LAM adaptive (LAMA) properties of estimates based on a sharper inequality than that in Hájek's (1972) Theorem 4.1, constructed LAM estimates under general conditions and gave sufficient conditions for the LAM and LAMA properties. We shall use these results here. The application is easy, but somewhat complicated by the fact that stochastic approximation involves a choice of design (choosing points at which observations are made) and thus is not an estimation problem only.

[1)] Research begun during author's visit at the University of Bern in 1979 and partially supported by NSF grants MCS-7802846 and MCS-8102233.

1.3. Remark. Related results. (i) Tusnády (1974) proved that a RM method yields estimates with a property we show in Lemma 3.2 for the adaptive RM method.

(ii) In a special case, the adaptive RM method yields a recursive LAMA estimate for a location parameter problem. Beran's (1978) (non-recursive) adaptive estimate for such a problem requires less assumptions and is also LAMA.

(iii) Fabian (1980, unpublished) contains the results reported here, but obtained only for f linear; but contains also a result that the recursive estimate proposed by Sakrison (1965, 1966) and modified in Fabian (1978) is LAM.

2. The assumptions.

We shall state basic assumptions concerning the functions f_δ the zero point δ of which is sought, the densities considered and describe the adaptive RM method.

2.1. Assumption. f is a function defined on R and for every $\varepsilon > 0$

(1) $$\sup\{f(x);\ -\varepsilon^{-1} < x < -\varepsilon\} < 0, \quad \inf\{f(x);\ \varepsilon < x < \varepsilon^{-1}\} > 0,$$

f is bounded on bounded intervals, has a uniformly continuous derivative f' and

(2) $$f'(0) = d > 0.$$

2.2. Notation. R denotes the real line, R^k the k-dimensional Euclidean space.

ι is the identity function on R, for each $\delta \in R$, f_δ denotes $f(\iota - \delta)$, i.e. $f_\delta(x) = f(x - \delta)$ for all x.

$\langle\Omega,\underline{\Omega}\rangle$ is a measurable space. If P, P_δ etc. denote probabilities on $\underline{\Omega}$, E, E_δ etc. denote the corresponding expectations. If X is a random variable PX^{-1} is the probability distribution of X and $\int\cdot dPX^{-1}$ is denoted by E^X. η is the expectation with respect to the standard normal probability distribution, $\Rightarrow$ indicates vague convergence.

λ is the Lebesgue measure on R. If h is a function on a subset $\mathcal{D}h$ of R^k and into $L_2(\lambda)$, $0 \in (\mathcal{D}h)^\circ$, then h has a differential $\dot{h}$ in $L_2(\lambda)$ at 0 if the function e defined by

$$e(x) = \frac{1}{\|x\|}[h(x) - h(0) - x'\dot{h}]$$

has a limit (in $L_2(\lambda)$) 0 at 0.

G is the family of all densities g with respect to λ which are symmetric ($g(t) = g(-t)$ for all $t \in R$), non-increasing on $[0,\infty)$, admit a continuous derivative g' on R, and for which $0 < I(g) = \int(g'/g)^2 g\,dt < \infty$ and $\int|t|g(t)dt<\infty$.

2.3. Assumption. Assumption 1 holds. $\underline{\Omega}$ is a σ-algebra, X_n, Y_n, U_{n+}, U_{n-} are random variables, $\langle m_\ell\rangle$ is an increasing sequence of integers such that $m_\ell/\ell \to 0$ as $\ell \to \infty$. For each n, $\mathcal{F}_n = \sigma(\{Y_i;\ i < n\} \cup \{U_{\ell+};\ m_\ell < n\} \cup \{U_{\ell-};\ m_\ell < n\}$ is the σ-algebra generated by the random variables indicated.

The random variables satisfy

(1) $$X_n = X_{n-1} - n^{-1}D_n h_n(Y_n) - n^{-1}(\log n)^{-1+\varepsilon}\tilde{Y}_n,\ X_0 = 0,$$

with an $\varepsilon \in (0,\frac{1}{6})$ (we achieve a simplification by denoting X_{n+1} in F by X_n here) and with h_n estimates of $-g'/g$, D_n estimates of $(I(g)d)^{-1}$ and $\tilde{Y}_n$ truncations of Y_n, all as described in F §4.2, U_ℓ functions of $U_{\ell+}, U_{\ell-}$.

For each $\langle\delta,g\rangle \in R \times G$, $E_{\delta,g}$ is an expectation on $\underline{\Omega}$ under which, for each $n = 1,2,\ldots$, Y_n has a conditional - given $\mathcal{J}_n$ - density $g(\iota - f_\delta(X_{n-1}))$. If $n = m_\ell$ then U_{n+} has a conditional density $g(\iota - f_\delta(X_{n-1}) + c_n)$ given $\mathcal{J}_n$, U_n and U_{n-} has a conditional density $g(\iota - f_\delta(X_{n-1}) - c_n)$ given $\mathcal{J}_n$, Y_n, U_{n+}, with c_n numbers, $c_n \to 0$.

2.4. Condition. φ is a one-to-one function with the domain $\mathcal{D}\varphi$ a subset of R^k for a k, 0 is an interior point of $\mathcal{D}\varphi$, the range $\mathcal{R}\varphi$ of φ is a subset of G. The function h defined on $R \times \mathcal{D}\varphi$ by $h(\delta,u) = \sqrt{\gamma(\iota-\delta)}$, $\gamma = \varphi(u)$ has a differential $\dot{h}$ at 0 in $L_2(\lambda)$.

3. Results.

3.1. Lemma. Let E be an expectation on $\underline{\Omega}$. Suppose $\langle\mathcal{J}_n\rangle$ is a non-decreasing sequence of sub-σ-algebras of $\underline{\Omega}$, X_n, V_n, W_n, Γ_n, Φ_n random variables. Suppose Γ_n, V_{n-1}, W_{n-1} are $\mathcal{J}_n$-measurable,

(1) $$\Gamma_n \to \underline{1} \quad \text{with prob. 1,}$$

(2) $$E_{\mathcal{J}_n} V_n = 0,\ \limsup |E_{\mathcal{J}_n} V_n V_n'| \leqq C \in R,$$

(3) $$E_{\mathcal{J}_n} W_n = 0,\ E_{\mathcal{J}_n}(V_n - W_n)^2 \to 0 \quad \text{w.p. 1,}$$

(4) $$X_n = (1 - n^{-1}\Gamma_n)X_{n-1} + n^{-1}V_n,$$

(5) $$S_n = \frac{1}{n}\sum_{j=1}^{n} W_j .$$

Then

(6) $$n^{1/2}[S_n - X_n] \to 0 \quad \text{in prob.}$$

Proof. Set $U_1 = 0$, $U_n = (1 - n^{-1})U_{n-1} + n^{-1}V_n$. It follows that

(7) $$n^{1/2}(X_n - U_n) \to 0 \quad \text{in prob.}$$

This can be shown by an adaptation of the proof in Fabian (1968). Similar adaptations were used by Berger (1979) and Pantel (1979) and a paper in preparation by Berger (hopefully to appear in Zeitsch. f. Wahrsch. u. v. Geb.) will make the derivation of (7) immediate. Because (7) and since $U_n = \frac{1}{n}\Sigma_{i=1}^n V_i$, it is enough to prove that, with $Z_i = V_i - W_i$,

(8) $$\frac{1}{\sqrt{n}}\sum_{i=1}^{n} Z_i \to 0 \quad \text{in prob.}$$

But that obtains from (3) and the Chebyshev-like inequality of Dubins and Savage (1965) which gives, for every $b > 0$

$$P\{n^{-1/2}|\sum_{j=1}^{n} Z_j| > b^{-2}\frac{1}{n}\sum_{j=1}^{n} E_{\mathcal{F}_j} Z_j^2 + b\} \leq 2b.$$

3.2. Lemma. Let Assumption 2.3 hold, $<\theta,g> \in R \times G$, let $I_n = nd^2 I(g)$, ,

(1) $$\gamma_n = -I_n^{-1/2}\sum_{i=1}^{n}\frac{g'}{g}(Y_i - f_\theta(X_{i-1})).$$

Then

(2) $$I_n^{1/2}(X_n - \theta) - \gamma_n \to 0 \quad \text{in } E_{\theta,g}\text{-prob.}$$

Proof. Theorem F.4.1 applies and we shall borrow from the proof. The last relation in the proof of Theorem F.3.1 is (assuming d=1, θ=0 and denoting $E_{0,g}$ by E) is

(3) $$X_n = (1 - n^{-1}\Gamma_n)X_{n-1} + n^{-1}V_n, \quad \Gamma_n \to 1,$$

where Γ_n are $\mathcal{F}_n$-measurable random variables,

(4) $$U_n = -n^{-1}D_n h_n(Y_n) - n^{-1}(\log n)^{-1+\varepsilon}\tilde{Y}_n$$

and

(5) $$V_n = n[U_n - N_n],\quad N_n = E_{\mathcal{F}_n}U_n$$

(see the line preceding (F.3.1.4) and the beginning of the paragraph containing (F.3.1.6); in F, V_n equal to our $-V_n$ is used).

By (F.3.24), $E_{\mathcal{F}_n}V_n^2 \to I(g)^{-1}$ and thus (3.1.2) holds. Relations (3.1.1) and (3.1.4) are satisfied.

Define

(6) $$W_n = -\frac{1}{I(g)}\frac{g'}{g}(Y_n - f_\theta(X_{n-1})).$$

We obtain $$0 = \int(g'/g)g\,d\lambda \in E_{\mathcal{F}_n}W_n.$$

Next,

$$E_{\mathcal{F}_n}(V_n - W_n)^2 = \mathrm{Var}_{\mathcal{F}_n}(V_n - W_n) \leq 2E_{\mathcal{F}_n}[D_n h_n(Y_n) - W_n]^2 + 2\mathrm{Var}_{\mathcal{F}_n}(\log n)^{-1+\varepsilon}\tilde{Y}_n.$$

This last term converges to 0 by (F.3.23). The term which preceeds is not greater than

(7) $$4D_n^2 E_{\mathcal{F}_n}(h_n(Y_n) - I(g)W_n)^2 + 4(D_n I(g) - 1)^2 E_{\mathcal{F}_n}W_n^2.$$

The second term here goes to 0 since $D_n \to I^{-1}(g)$ by Assumption (F.2.4) and $E_{\mathcal{F}_n}W_n^2 = 1$.

The function $\omega \rightsquigarrow \int[h_n(t + f_\theta(X_{n-1}(\omega)) + \frac{g'}{g}(t)]^2 dG(t)$ converges to 0 as $n \to \infty$ by Assumption (F.2.5). Since it is a version of the conditional expectation in the first term in (7), that term converges to 0, and $E_{\mathcal{J}_n}(\bar{V}_n - W_n)^2 \to 0$.

We have verified the conditions of Lemma 3.1 which now implies (2).

3.3. Theorem. Let Assumption 2.3 hold, $\theta \in R$, let φ satisfy Condition 2.4, $g = \varphi(0)$, $I_n = nd^2 I(g)$.

Then

(i) If $\delta_n \in R$, $u_n \in \mathcal{D}\varphi$, and if $\langle\sqrt{n}\,(\delta_n - \theta)\rangle$, $\langle\sqrt{n}\,|u_n|\rangle$ are bounded sequences then

$$(1)\qquad E_{\delta_n,\varphi(u_n)}^{\sqrt{I_n}\,(X_n-\delta_n)} \Rightarrow \eta\,.$$

(ii) If $c \in R$, $\ell = 1 - \chi_{(-c,c)}$ then

$$(2)\qquad \lim_{K\to\infty} \lim_n \sup_{\substack{|\delta-\theta|+\\ \|u_n\|\le Kn^{-1/2}}} E_{\delta,\varphi(u_n)}\, \ell(\sqrt{I_n}\,(X_n - \delta)) = \eta\ell$$

and for every sequence of estimates ξ_n, with $\mathcal{J}_{n+1}$-measurable ξ_n,

$$(3)\qquad \lim_{K\to\infty} \lim_n \inf \sup_{|\delta-\theta|\le Kn^{-1/2}} E_\delta \ell(\sqrt{I_n}\,(\xi_n - \delta)) \geq \eta\ell\,.$$

Proof. Assume $d = 1$ without loss of generality. Let $E_{n,\delta,\varphi(u)}$ denote the restriction of the expectation $E_{\delta,\varphi(u)}$ to $\mathcal{J}_{n+1}$-measurable functions, $\mathcal{E}_\varphi = \langle E_{n,\delta,\varphi(u)};\ \delta \in R,\ u \in \mathcal{D}\varphi,\ n = 1,\ldots\rangle$, $\mathcal{E}_0 = \langle E_{n,\delta,g};\ \delta \in R,\ n = 1,\ldots\rangle$ with $g = \varphi(0)$. The family $\mathcal{E}_\varphi$ satisfies Condition LAN $\langle\theta,0\rangle$, nM, $\tilde{\gamma}_n\rangle$ with suitable $\tilde{\gamma}_n$ and $M = 4\int(\dot{h})(\dot{h})'d\lambda$ (cf. Definition FH.2.2). This has been shown in Example 3.11 of Fabian and Hannan (1982b) illustrating the use of sufficient conditions for the LAN property in case of dependent random variables.

Also, $\mathcal{E}_0$ satisfies Condition LAN$\langle\theta$, $nI(g)$, $\gamma_n\rangle$ with γ_n given by (3.2.1). Hence the second part of assertion (i) follows by Theorem FH.2.6.

The first coordinate $\dot{h}^{(1)}$ of $\dot{h}$ satisfies $\dot{h}^{(1)} = -\frac{1}{2}\,g'/\sqrt{g}$ a.e. (λ). For $j \neq 1$, $\dot{h}^{(j)}$ is a limit in $L_2(\lambda)$ of functions of form $c(\sqrt{g_1} - \sqrt{g}\,)$ with $g_1 \in G$. Thus $\dot{h}^{(j)}$ is symmetric around 0 (a.e. and $\dot{h}^{(1)}$ is odd around 0 and $M^{(1,j)} = \int \dot{h}^{(1)}\dot{h}^{(j)}d\lambda = 0$.

$\mathcal{E}_0$ is called a singular subproblem in Definition FH.7.3. The function φ specifies the subproblem $\mathcal{E}_\varphi$ (formally, in Definition FH.7.3, $\langle\delta,u\rangle \rightsquigarrow \langle\delta,\varphi(u)\rangle$ is called a subproblem), both $\mathcal{E}_0$, $\mathcal{E}_\varphi$, are LAN, and $\mathcal{E}_\varphi$ satisfies Condition FH.7.8 of $M^{(1,j)} = 0$ for $j \neq 1$. Lemma 3.2 asserts that $\langle X_n\rangle$ is a regular sequence of estimates for the singular problem (cf. Def. FH.6.2). With $\mathcal{Q}$ the family of all subproblems $\mathcal{E}_\varphi$, Theorem FH.7.10 applies and asserts that $\langle X_n\rangle$ is an LAMA sequence of estimates, and this implies (2). Relation (1) follows by Theorem FH.6.3.

3.4. Remark. Optimality. Part (ii) of the preceding theorem shows that the LAM risk (the left hand side of (3.3.3)) of any sequence of $\mathcal{F}_{n+1}$-measurable estimates ξ_n is at least as large as the LAM risk for $\langle X_n \rangle$. The complication is that the $\mathcal{F}_n$ depend on the choice of $\langle X_n \rangle$. This complication is without consequence if $f = d\iota$, but is substantial for some other f.

Consider the case $f = d\iota$ with d known, U_ℓ constants, set $Z_j = Y_j - dX_{j-1}$. Then $\mathcal{F}_{n+1}$ is equal to the σ-algebra generated by $\{Z_1,\ldots,Z_n\}$ with Z_i independent and with density $g(\iota + d\delta)$ under $F_{n\delta}$. It follows that $E_{n\delta}$ does not depend on the choice of $X_1, X_2, \ldots$, and (3.3.3) is not affected by a possible change in this choice.

It follows that $\langle X_n \rangle$ is optimal in minimizing the LAM risk in case $f = \delta\iota$; it would stay optimal if judged by the supremum of the LAM risk over a family $\mathcal{F}$ of functions f which includes $d\iota$ for a $d > 0$. A case of $f \neq d\iota$ is considered next.

3.5. Example. Let f be as in Assumption 2.1 with $d = 1$, and let $\tilde{f} = f \circ (\iota - 3) - 3$ satisfies Assumption 2.1 with $d = 2$. Then applying the adaptive RM method to f_δ we obtain (3.3.2) with $d = 1$. Applying the method to $\tilde{f}_\delta$ gives $\tilde{X}_n$ with $X_n = \tilde{X}_n - 3$ satisfying (3.3.2) with $d = 2$. It is simply more efficient, in this case, to estimate the point at which f is equal to 3 than that at which it is 0.

REFERENCES

Abdelhamid, S.N. (1973). Transformation of observations in stochastic approximation. Ann. Statist. 1 1158-1174.

Anbar, D. (1973). On optimal estimation methods using stochastic approximation procedures. Ann. Statist. 1 1175-1184.

Beran, R. (1978). An efficient and robust adaptive estimator of location. Ann. Statist. 6 292-313.

Berger, E. (1979). A note on the invariance principle for stochastic approximation procedures in a Hilbert space. Preprint. Institut für Mathematische Statistics, Universität Göttingen.

Dubins, L.E. and L.J. Savage (1965). A Tchebycheff-like inequality for stochastic processes. Proc. Nat. Acad. Sci. U.S.A. 53 274-275.

Fabian, V. (1968). On asymptotic normality in stochastic approximation. Ann. Math. Statist. 39 1327-1332.

Fabian, V. (1973). Asymptotically efficient stochastic approximation; the RM case. Ann. Statist. 1 486-495.

Fabian, V. (1978). On asymptotically efficient recursive estimation. Ann. Math. Statist. 6 854-866.

Fabian, V. and J. Hannan (1982a). On estimation and adaptive estimation for locally asymptotically normal families. Z. Wahrscheinlichkeitstheorie verw. Gebiete 59 459-478.

Fabian, V. and J. Hannan (1982b). Sufficient conditions for local asymptotic normality. Manuscript, a slight revision of RM-403, Dept. of Statistics and Probability, Michigan State University, 1980.

Hájek, J. (1972). Local asymptotic minimax and admissibility in estimation. Proc. 6. Berkeley Symp. Math. Statistics and Probability, Univ. Calif. Press, 172-194.

Pantel, M. (1979). Adaptive Verfahren der Stochastischen Approximation. Dissertation. Universität Essen-Gesamthochschule, Jan. 1979.

Robbins, H. and Monro, S. (1951). A stochastic approximation method. Ann. Math. Statist. 22 400-407.

Sakrison, D.J. (1965). Efficient recursive estimation; application to estimating the parameters of a covariance function. Internat. J. Engrg. Sci 3 461-483.

Sakrison, D.J. (1966). Stochastic approximation: A recursive method for solving regression problems. Advances in Communication Systems 2 51-106.

Tusnády, G. (1974). On the updated maximum likelihood estimators. Studia Scientiarum Mathematicarum Hungarica 9 377-389.

DYNAMIC ALLOCATION INDICES FOR BAYESIAN BANDITS

J.C. Gittins,
Mathematical Institute,
24-29, St. Giles, Oxford,
OX1 3LB, England.

ABSTRACT. Bather has recently obtained an asymptotic result for the dynamic allocation index of the bandit process defined by a Wiener reward process with unknown drift. This result is shown to apply to other bandit processes generated by a sequence of independent identically distributed random variables under reasonably general conditions. Computational results are presented for three such processes.

1. INTRODUCTION

In the early 1960's Lindley [15] and Chernoff [3] noted that sequential decision problems involving sequences of independent identically distributed normal random variables with unknown mean, and formulated in a Bayesian fashion , may be approximated by stopping problems in terms of Wiener processes. Such problems may be solved by finding the solution to a partial differential equation, known as the heat equation, with appropriate boundary conditions. Bather and Chernoff (for example see Bather [1] and Chernoff [4]) obtained approximate solutions to a variety of interesting problems in this way.

Some years later the present author in a paper with Jones [8] showed that the optimal policy for the discounted Bayesian multi-armed bandit may be described in terms of an index, called a dynamic allocation index, defined for each arm as a real function of its history. The optimal policy is at each stage to pull an arm for which the index value is maximal in the set of available arms. This result holds when the reward processes associated with the various arms form quite general random sequences, and extends to a variety of stochastic scheduling and search problems. These applications are reviewed in two more recent papers, [6] and [7]. Important contributions to the theory of dynamic allocation indices have been made by Glazebrook ([11] and earlier papers), Kelly [13], Nash (in particular [17]), and Whittle [18 and 19].

A Wiener process with unknown drift μ and variance ψ^2 per unit time defines an arm of a multi-armed bandit, or bandit process. if the increments are regarded as rewards. A two-armed bandit with one arm like this, and the other a standard arm producing rewards at a known and constant rate, reduces to a stopping problem whose solution yields the dynamic allocation index for the non-standard arm. Using the methods set out in [1] Bather has recently [2] shown that if μ has a $N(w,v)$

prior distribution, and rewards at time t are discounted by the factor $e^{-\alpha t}$, the index

$$\nu(w,v) = w+\alpha^{\frac{1}{2}}\tau u(\frac{v}{\alpha\tau^2}) , \qquad (1)$$

where the function u(y) is positive, increasing, and such that $y^{-1}u(y)$ is bounded above by $2^{-\frac{1}{2}}$ and tends to $2^{-\frac{1}{2}}$ ($u(y) \tilde{<} 2^{-\frac{1}{2}}y$) as y tends to zero.

This limit leads to a similar one for a bandit process generating a sequence of normally distributed rewards. It will be shown that an initial assumption of normality is not necessary, and the argument may be extended to the seemingly rather different situation where the object is to minimise the expected number of pulls before a value is sampled which lies in some target range. This paper also reports on a computational investigation of these asymptotic results for the cases of Bernoulli and negative exponential bandit processes. Bather's analysis turns out to suggest the form of a useful approximation even for values of the sufficient statistics well outside the region in which asymptotic results might be expected to apply.

As a by-product of these investigations we also obtain an apparently new result in the theory of scheduling jobs of random length on a single machine, when the machine can switch from job to job in negligible time, whether or not it has just completed processing a job. On the assumptions that rewards accrue when jobs are completed, are exponentially discounted according to the completion times, and are the same for every job, the index policy which maximises the total expected reward turns out also to maximise the expected reward from the first job to be completed.

2. REVIEW OF SOME PROPERTIES OF BANDIT PROCESSES

For a bandit process D which is initially in state x let $R_\tau(x)$ be the expected discounted reward if it is continued without interruption up to the stopping time τ, which typically depends on the history of the process up to that time. Let

$$W_\tau(x) = E\int_0^\tau e^{-\alpha t}dt = \alpha^{-1}(1-Ee^{-\alpha\tau}).$$

The dynamic allocation index for D in state x may be written as

$$\nu(x) = \sup_{\tau>0} \{R_\tau(x)/W_\tau(x)\}, \qquad (2)$$

and may thus be regarded as the rate at which rewards are generated by D up to a time which is chosen so as to maximise this rate.

For a discrete-time process in which the n'th reward is discounted by the factor a^{n-1} $(0 < a < 1)$ $W_\tau(x)$ is defined as $E\,\Sigma_{n=1}^{\tau} a^{n-1} = (1-a)^{-1}(1-Ea^{\tau})$. This differs from the definition just given by the discrete time correction factor $\alpha(1-e^{-\alpha})^{-1}$ if a is set equal to $e^{-\alpha}$.

A Wiener bandit process of the type considered by Bather is a standard bandit process with parameter μ_0 if the drift is known to take the value μ_0. A discrete-time bandit process generated by a sequence of independently and identically distributed rewards is a standard bandi process with parameter θ_0 if the expectation of each reward is known to be θ_0. In either case rewards are generated at the uniform rate given by the parameter value, which is therefore the value of the dynamic allocation index for such a process. For the continuous-time case this is confirmed by setting $w = \mu_0$ and $v = 0$ in equation (1).

If the expected value θ of successive rewards, or the drift μ of the Wiener reward process, is known only in terms of a prior distribution Π, then Π is itself the state of the bandit process and

$$\nu(\Pi) \geq E(\theta/\Pi)[\text{or } E(\mu/\Pi)]. \tag{3}$$

This follows from the definition (2) when we note that $R_\tau(\Pi)/W_\tau(\Pi) = E(\theta/\Pi)[\text{or } E(\mu/\Pi)]$ for any τ which does not depend on the history of the process. In general (3) holds with strict inequality as we can increase $R_\tau(\Pi)/W_\tau(\Pi)$ by making τ larger or smaller according as the rate of observed rewards is high or low. The second term on the right hand side of (1) represents the extent of the resulting increase in reward rate for a Wiener bandit process.

As successive rewards are obtained the prior distribution Π is modified by Bayes theorem to form posterior distributions. These constitute the sequence of states through which the bandit process passes. The continuous sequence of distributions generated in this way for the drift of a Wiener bandit process are all normal distribution if the prior distribution is normal, as is easily verified, so that an equation of the form (1) holds throughout. This property, known as *closure under sampling*, of the family of normal prior distributions means that the state of the bandit process is defined by the parameters of the current posterior distribution.

3. THE NORMAL REWARD PROCESS (KNOWN VARIANCE)

Each of the specific bandit processes considered in this paper is generated by a sequence of independent and identically distributed random variables $(X_i; i=1,2,\ldots)$. Except for the last example, these random variables themselves constitute the reward sequence when the bandit process is continued, and the n'th reward in any sequence is discounted by the factor a^{n-1} $(0 < a < 1)$. We first suppose the X_i's to be normally distributed with mean θ and known variance σ^2. No detailed calculations for this case are available, but it is an important intermediate stage between the Weiner bandit process and the other processes to be considered later.

This normal reward process may be regarded as generated by a Wiener process as in the case considered by Bather, but with observations and decision times at unit intervals rather than continuously. To establish this equivalence note that at the end of a unit time interval a Wiener bandit process with drift μ, variance τ^2 per unit time, and discount parameter α yields a $N(\theta,\sigma^2)$ reward with

$$\theta = \int_0^1 \mu e^{-\alpha t} dt = \mu\alpha^{-1}(1-e^{-\alpha}) \text{ and } \sigma^2 = \int_0^1 \tau^2 e^{-\alpha t} dt = \tau^2\alpha^{-1}(1-e^{-\alpha}).$$

The discount factor per unit time is $e^{-\alpha}$, which we therefore set equal to a, and if the prior distribution for μ is $N(w,v)$ then θ also has a normal prior distribution with mean $w\alpha^{-1}(1-e^{-\alpha})$ and variance $v\alpha^{-2}(1-e^{-\alpha})^2$.

For the normal process with a normal prior distribution for θ it is convenient to express this prior distribution in the form $N(\bar{x},\sigma^2 n^{-1})$. This is the posterior distribution for θ after n observations with mean value $\bar{x}$ when the prior is the improper uniform distribution on the entire real line. Writing the prior distribution in this way amounts to thinking in terms of n *hypothetical* observations with mean $\bar{x}$. The advantage of this convention is that as actual rewards are generated by the process the posterior distribution is always of the same form if we simply treat hypothetical and actual observations in the same way. Thus the state of the bandit process is defined by the parameters $\bar{x}$ and n, where $\bar{x}$ and $\sigma^2 n^{-1}$ are the mean and variance of the current distribution for θ, and n is not restricted to integer values.

As compared with the Wiener bandit process with the same initial state, the effect of restricting observation and decision times to the integers is clearly to reduce the dynamic allocation index. Making the appropriate substitutions in (1) and applying the discrete time correction factor $\alpha(1-e^{-\alpha})^{-1}$, the normal process dynamic allocation index $\nu_a(\bar{x}, n)$ satisfies the right hand inequality below.

$$0 < \nu_a(\bar{x},\ n) - \bar{x} \stackrel{\sim}{<} (1-a)^{\frac{1}{2}}\sigma\ u(\frac{1}{n(1-a)}). \tag{4}$$

The left hand inequality is a particular case of (3).

Simple location and scale invariance arguments show that

$$\nu_a(\bar{x},\ n) = \bar{x} + \sigma g(a,n) \tag{5}$$

for some function $g(a,n)$. From (4) and its derivation it follows that

$$g(a,n) \stackrel{\sim}{<} (1-a)^{\frac{1}{2}}\ u([n-na]^{-1}),$$

with approximate equality when the decision times are close together in an appropriate sense. If we consider the normal process approximation to a Wiener bandit process which is obtained by setting an interval T between decision times, we have $a = \exp(-\alpha T)$, while at a given point in the process $n\alpha T(\simeq n-na)$ remains fixed as T and n are varied. Thus, for a given value of $n(1-a)$ the approximation should improve, and in the limit be correct, as T decreases to 0, which is to say as a approaches 1. Since $y^{-1}\ u(y) \stackrel{\sim}{<} 2^{-\frac{1}{2}}$ as $y \to 0$ it follows that

$$[\nu_a(\bar{x},n) - \bar{x}]\sigma^{-1}\ n(1-a)^{\frac{1}{2}} \stackrel{\sim}{<} 2^{-\frac{1}{2}} \text{ as } n(1-a) \to \infty \text{ and } a \to 1. \tag{6}$$

For this normal reward process $\bar{x}$ and $\sigma^2 n^{-1}$ are the mean and variance of the current distribution for θ, the expected reward. Provided these quantities take a similar form the equation (5) and asymptotic result (6) remain approximately valid for large n for non-normal reward processes under suitable conditions. For any reward process the dynamic allocation index depends only on the random process defined by the mean of the current distribution for the expected reward. Thus the required conditions are ones which ensure that the change in $\bar{x}$ as n increases to $n + m$ has a distribution which for large n is close to its distribution for the normal case. The theorem of the following section gives us most of what we need for present purposes.

4. ASYMPTOTIC VALIDITY OF THE NORMAL CASE

The general setup is the standard one for sequential Bayesian inference. The parameter space $\Theta(\ni\theta)$ is a Borel subset of R^r. A prior probability distribution Π_0 is defined on the Borel subsets of Θ, with a density π_0 with respect to some sigma-finite measure ζ. The observation space $X(\ni x)$ is a Borel subset of R. A conditional probability distribution function L_θ (the likelihood function) is defined on the Borel subsets of X for every $\theta \in \Theta$. $L_\theta(A)$ is a Baire function on Θ for every Borel set $A \subset X$ and has a density ℓ_θ with respect to some sigma-finite measure μ. The mean θ_1 and variance θ_2 of L_θ are both assumed to exist.

Independent observations $X_1, X_2, \ldots$ all with the distribution L_θ are made sequentially, and generate the probability measure P_θ on R^∞. Particular values of the X_i's, such as observations which have already been made, are indicated by lower case letters; $h_n = (x_1, x_2, \ldots, x_n)$. The generalised Bayes theorem tells us that a version Π_n of the distribution of θ conditional on h_n (called the posterior distribution of θ after n observations) exists, having the density $\pi_n(\theta) \propto \pi_0(\theta)\Pi_{i=1}^n \ell_\theta(x_i)$ with respect to ζ. P_n is the probability measure over $\Theta \times R^\infty$ defined by Π_n and P_θ.

Our key asymptotic result involves the following random variables which, together with θ_1 and θ_2, are defined in terms of P_n; s is a positive constant, $\bar{x}_n = n^{-1}(x_1+x_2+\ldots+x_n)$.

$$U = m^{-\frac{1}{2}}s^{-1}\left(\sum_{i=n+1}^{n+m} X_i - m\theta_1\right), \quad U' = m^{-\frac{1}{2}}\theta_2^{-\frac{1}{2}}\left(\sum_{i=n+1}^{n+m} X_i - m\theta_1\right),$$

$$V = n^{\frac{1}{2}}s^{-1}(\theta_1 - \bar{x}_n), \qquad V' = n^{\frac{1}{2}}\theta_2^{-\frac{1}{2}}(\theta_1 - \bar{x}_n),$$

$$W = (m^{-1}+n^{-1})^{-\frac{1}{2}}\,(m^{-\frac{1}{2}}U + n^{-\frac{1}{2}}V).$$

F_U, $F_{U'}$, F_V, $F_{V'}$ and F_W are their distribution functions; F_1 and F_2 are those of θ_1 and θ_2. Conditional distribution functions are denoted by $F_{U'|\theta}$, $F_{2|V'}$ etc. Φ and ϕ are the standard normal distribution and density functions.

The following conditions are required.

(i) There is a sequence of Baire functions s_n defined on $h_n (n=0,1,2,\ldots)$ such that, given ϵ and s (both > 0), $\Pi_n(|\theta_2 - s^2| > \epsilon) \to 0$ as $n \to \infty$ for any sequence $h_n (n=0,1,2,\ldots)$ on which $s_n(h_n)$ takes the fixed value s for every n. A sequence of h_n's with this property will be termed a *fixed s sequence*.

(ii) $F_{V'}(v) \to \Phi(v)$ on any fixed s sequence, $v \in R$.

(iii) For any bounded interval $I \subset R$ and $\epsilon > 0$,

$$\Pi_n(|F_{U'|\theta}(u) - \Phi(u)| < \epsilon,\ u \in I) \to 1$$

on any fixed s sequence as m and $n \to \infty$ in such a fashion that $mn^{-1} \geq$ some fixed positive c.

Note that the Lindberg-Levy central limit theorem tells us that $F_{U'|\theta}(u) \to \Phi(u)$ uniformly in $u \in R$ as $m \to \infty$. Condition (iii) is a related, but not equivalent, property.

When the X_i's are a reward sequence and $E\theta_1 = \bar{x}_n$ our interest, as remarked at the end of section 3, is in showing that for large n the difference $\Delta\bar{x} = \bar{x}_{n+m} - \bar{x}_n$ has a distribution close to its distribution for the normal case. This is the meaning of the theorem which follows. Note that $\Delta\bar{x} = sm^{\frac{1}{2}}n^{-\frac{1}{2}}(m+n)^{-\frac{1}{2}}W$.

Theorem 1 (Normal Limit Theorem). 1. If L_θ is $N(\theta_1,\sigma^2)$, where $\sigma^2(=s^2)$ is known, and Π_0 is the improper uniform distribution on the entire real line, then $F_W(w) = \Phi(w)$, $w \in R$. 2. Under conditions (i), (ii) and (iii), $F_W(w) \to \Phi(w)$, $w \in R$, on any fixed s sequence as m and $n \to \infty$ in such a fashion that $mn^{-1} \geq$ some fixed positive c.

Proof. 1. $F_{U|\theta} = \Phi$, $\theta \in \Theta$. Thus $F_U = \Phi$ and U is independent of V. Also $F_V = \Phi$, since Π_n is $N(\bar{x}_n,\sigma^2 n^{-1})$. Part 1 of the theorem follows, since W is a linear combination of U and V with coefficients the sum of whose squares is one. 2. For simplicity of presentation, and with only an easily removable loss of generality, out proof is for the case $mn^{-1} = c$.

$$F_W(w) = \int F_{U|\theta}[(1+c^2)^{\frac{1}{2}}w - cn^{\frac{1}{2}}s^{-1}(\theta_1-\bar{x}_n)]d\Pi_n(\theta) . \tag{7}$$

For the case of part 1 this reduces to

$$F_W(w) = \int_{-\infty}^{\infty} \Phi[(1+c^2)^{\frac{1}{2}}w - cv]\phi(v)dv, \tag{8}$$

which, as just shown, is equal to $\Phi(w)$. The proof of part 2 consists of using conditions (i), (ii) and (iii) to obtain bounds on the difference between expressions (7) and (8).

We first show that, given $\epsilon > 0$, M may be chosen so that

$$|F_W(w) - \int_{-\infty}^{\infty} \Phi[(1+c^2)^{\frac{1}{2}}w - cv]dF_{V'}(v)| < 3\epsilon/5, \quad n > M. \tag{9}$$

Let $\xi = \Phi^{-1}(1-\epsilon/10)$, and $\delta(0<\delta<\frac{1}{2}s)$ be such that

$$|\Phi[(1+c^2)^{\frac{1}{2}}ws\theta_2^{-\frac{1}{2}} - cv] - \Phi[(1+c^2)^{\frac{1}{2}}w - cv]| < \epsilon/5$$

if $|s-\theta_2^{\frac{1}{2}}| < \delta$ and $|v| < \xi$. There must be such a δ because of the uniform continuity of continuous functions on compact intervals. Now choose M so that

$$\Pi_n(\{|V'|\geq\xi\} \cup \{|\theta_2-s^2|\geq\delta\} \cup \{|F_{U'|\theta}(u)-\Phi(u)| \geq \epsilon/5 \text{ for some } |u|<2(1+c^2)^{\frac{1}{2}}|w|+c\xi\}) < \epsilon/5, \quad n > M. \tag{10}$$

This is possible because of conditions (ii), (i) and (iii) in turn.

Next note that

$$F_{U|\theta}[(1+c^2)^{\frac{1}{2}}w-cn^{\frac{1}{2}}s^{-1}(\theta_1-\bar{x}_n)] = F_{U'|\theta}[(1+c^2)^{\frac{1}{2}}ws\theta_2^{-\frac{1}{2}}-cn^{\frac{1}{2}}\theta_2^{-\frac{1}{2}}(\theta_1-\bar{x}_n)] .$$

It thus follows from (10) that the integrand in (7) may be replaced by $\Phi[(1+c^2)^{\frac{1}{2}}ws\theta_2^{-\frac{1}{2}}-cn^{\frac{1}{2}}\theta_2^{-\frac{1}{2}}(\theta_1-\bar{x}_n)]$, and then by $\Phi[(1+c^2)^{\frac{1}{2}}w-cn^{\frac{1}{2}}\theta_2^{-\frac{1}{2}}(\theta_1-\bar{x}_n)]$, with an error at each stage of no more than $\epsilon/5$, except on a set of Π_n measure at most $\epsilon/5$, on which the total error of the two changes is at most one. Now writing $dF_{2|V'}(\theta_2)dF_{V'}(v)$ in place of $d\Pi_n(\theta)$, which we can do since the integrand now depends on θ only through θ_2 and V', and carrying out the integration with respect to θ_2, the error bound (9) follows.

The rest of the proof is along similar lines. The convolution $\int \Phi[(1+c^2)^{\frac{1}{2}}w-cv]dF_{V'}(v)$ may be rewritten as $\int F_{V'}[(1+c^2)^{\frac{1}{2}}w-c^{-1}y]\phi(y)dy$. Choose N so that $|F_{V'}(v)-\Phi(v)| < \epsilon/5$ for $|v| \le (1+c^{-2})^{\frac{1}{2}}|w|+c^{-1}\xi$ and $n > N$, which is possible by (ii). Arguing along the lines of the previous paragraph it follows that

$$|\int_{\infty}^{\infty}\Phi[(1+c^2)^{\frac{1}{2}}w-cv]dF_{V'}(v) - \int_{\infty}^{\infty}\Phi[(1+c^{-2})^{\frac{1}{2}}w-c^{-1}y]\phi(y)dy| < 2\epsilon/5,\ n > N] \tag{11}$$

Since $\int \Phi[(1+c^2)^{\frac{1}{2}}w-c^{-1}y]\phi(y)dy = \Phi(w)$, we have from (9) and (11) that

$$|F_W(w)-\Phi(w)| < \epsilon ,\quad n > \max(M,N),$$

and the theorem is proved.

From the discussion of section 3 it follows that when for each Π_n the expected value of the next reward in the sequence is $\bar{x}_n$, and the conditions of the theorem hold, the asymptotic result (6) may be replaced by the limit

$$[\nu_a(\Pi_n,n)-\bar{x}_n]s^{-1}n(1-a)^{\frac{1}{2}} \to 2^{-\frac{1}{2}} \tag{12}$$

as $n(1-a) \to \infty$ on a fixed s sequence and $a \to 1$. The fact that condition (iii) requires $mn^{-1} \ge c$ might at first sight be expected to invalidate this conclusion, since the calculation of $\nu_a(\Pi_n,n)$ involves values of $\bar{x}_{n+m}$ for all $m > 0$. The reason it does not do so is that we are considering values of a close to 1. This means that many members of the sequence $\bar{x}_{n+m}(m = 0,1,2,...)$ have an appreciable effect in the calculation of $\nu_a(\Pi_n,n)$, and a many be chosen near enough to 1 for the higher values of m, for which the normal approximation holds, to swamp the effect of the lower values, for which it does not.

Applications of the theorem may be extended by considering a function $G(y,n)$ on $R^+ \times R$ satisfying appropriate regularity conditions. Suppose $G(y,n)$, $\partial G(y,n)/\partial y$, and $n^{\frac{1}{2}}[G(y,n+m) - G(y,n)]$, as n tends to infinity tend uniformly in bounded y intervals and $m \ge 0$ to $G(y)$,

$dG(y)/dy$, and 0, respectively, where $G(y)$ is continuously differentiable. Let $\Delta G = G(\bar{x}_{n+m}, n+m) - G(\bar{x}_n, n)$, and a *fixed $\bar{x}$ sequence* be a sequence of h_n's on which $\bar{x}_n = \bar{x}(n = 1,2,\ldots)$ for some $\bar{x}$.

Corollary 1.1. Under conditions (i), (ii) and (iii) and the above regularity conditions,

$$P_n(\{|\Delta G/\Delta\bar{x} - dG(\bar{x})/dy| > \epsilon\} \to 0$$

on any fixed $\bar{x}$ and s sequence, as m and $n \to \infty$ in such a fashion that $mn^{-1} \geq$ some fixed positive c.

Proof. For brevity write the conclusion of the corollary as $\Delta G/\Delta\bar{x} \overset{p}{\to} dG(\bar{x})/dy$. We have

$$\Delta G = G(\bar{x}+\Delta\bar{x}, n+m) - G(\bar{x}+\Delta\bar{x}, n) + \Delta\bar{x} \frac{\partial}{\partial y} G(\bar{x}+\alpha\Delta\bar{x}, n) ,$$

for some $\alpha \in (0,1)$. To establish the corollary it is therefore sufficient to show that (a) $\frac{\partial}{\partial y} G(\bar{x}+\alpha\Delta\bar{x}, n) \overset{p}{\to} \frac{dG(\bar{x})}{dy}$, and

$$\text{(b)} \quad (\Delta\bar{x})^{-1}[G(\bar{x}+\Delta\bar{x}, n+m) - G(\bar{x}+\Delta\bar{x}, n)] \overset{p}{\to} 0 .$$

The limit (a) follows from part 2 of the theorem, the uniform convergence of $\partial G(y,n)/\partial y$ to $dG(y)/dy$ on bounded y intervals, and continuity of dG/dy at $y = \bar{x}$. From part 2 of the theorem it also follows that

$$\lim_{\delta\to 0} [\liminf_{n\to\infty} P_n(n^{\frac{1}{2}}|\Delta\bar{x}| > \delta)] = 1. \tag{13}$$

Now writing the expression in (b) as $(n^{\frac{1}{2}}\Delta\bar{x})^{-1} n^{\frac{1}{2}}[G(\bar{x}+\Delta\bar{x}, n+m) - G(\bar{x}+\Delta\bar{x}, n)]$, the limit (b) follows from (13), the uniform convergence to zero of $n^{\frac{1}{2}}[G(y, n+m) - G(y,n)]$, and a further application of part 2 of the theorem.

This corollary may be applied to bandit processes for which the expected value of the next reward in the sequence, when the current distribution for θ is Π_n, is $G(\bar{x}_n, n)$, where $G(y,n)$ satisfies the stated conditions. A difference ΔG between members of the sequence of expected rewards is now well approximated for large n by multiplying the difference $\Delta\bar{x}$, for the situation where this sequence coincides with the $\bar{x}_n$'s, by $dG(\bar{x}_n)/dy$, which may be treated as independent of n provided we are considering only large values of n. Asymptotically, then, the relationship between the two expected reward sequences may be regarded as constant and linear. Under these circumstances it follows from the definition (2) that the same linear relationship must hold between the corresponding dynamic allocation indices. The asymptotic result corresponding to (12) is therefore

$$[\nu_a(\Pi_n,n) - G(\bar{x}_n,n)][s\, dG(\bar{x})/dy]^{-1} n(1-a)^{\frac{1}{2}} \to 2^{-\frac{1}{2}}$$

as $n(1-a) \to \infty$ on a fixed $\bar{x}$ and s sequence and $a \to 1$.

5. THREE MORE REWARD PROCESSES

In this section three bandit processes involving sequences of independent and identically distributed rewards are considered, each of them satisfying the conditions of the normal limit theorem. It is well-known, see for example Le Cam [14] and Lindley [16], that if $\Theta = R^r(\ni\theta=(\theta_1,\theta_2,\ldots,\theta_r))$ then under regularity conditions Π_n is asymptotically multivariate normal with a dispersion matrix whose (i,j) element is

$$-\sum_{k=1}^{n} \partial^2 \log \ell_\theta(x_k)/\partial\theta_i\partial\theta_j \;,$$

evaluated at the maximum likelihood value $\hat{\theta}_n$ of θ. This result certainly must remain true if $\hat{\theta}_n$ is replaced by the expected value of Π_n, thereby in principle provided a means of checking conditions (i) and (ii). However, it seems unlikely that the literature includes a limit theorem precisely of the required form. Rather than attempting to fill this gap we shall adopt the expedient of checking the conditions separately in each case. Some computational results are presented for the two cases for which these are available.

Normal Process (unknown variance)

X_i has a $N(\theta_1,\theta_2)$ distribution. $\Theta = \{(\theta_1,\theta_2) \in R \times R^+\}$. Improper prior density $\pi_0 \propto \theta_2^{-1}$. $s_n^2 = (n-1)^{-1}\Sigma(x_i-\bar{x}_n)^2$. Π_n is such that $n^{\frac{1}{2}}s_n^{-1}(\theta_1-\bar{x}_n)$ has a t distribution and $(n-1)s_n^2\theta_2^{-1}$ a chi-squared distribution, both with n-1 degrees of freedom.

Thus the current expectation of θ_1 is $\bar{x}_n$, and condition (i) of the theorem may be checked by applying Chebychev's theorem. Conditions (ii) and (iii) are both satisfied in an exact sense, without needing to take limits. The asymptotic result (12) therefore holds, and Π_n may be replaced by the sufficient statistics $\bar{x}_n$ and s_n^2. As yet no computational results are available for this case.

Bernoulli Process

X_i takes values 0 and 1 with probabilities $1-\theta_1$ and θ_1, $\theta_2 = \theta_1(1-\theta_1)$. Θ is the closed interval [0,1]. Improper prior density $\pi_0 \propto \theta_1^{-1}(1-\theta_1)^{-1}$. Let $\bar{x}_n = rn^{-1}$. Π_n has density $\pi_n \propto \theta_1^{r-1}(1-\theta_1)^{n-r-1}$, which is a beta density with parameters r and n-r, mean rn^{-1}, and variance $r(n-r)n^{-2}(n+1)^{-1}$.

Putting $s_n^2 = r(n-r)n^{-2}$, which means that a fixed s sequence is a fixed $\bar{x}$ sequence, condition (i) follows from Chebychev's theorem. Note that here a fixed s sequence defined for integer values of n involves non-integer values of r. Since the corresponding sequence of beta distributions Π_n is well-defined this causes no difficulty.

The distribution function for a standardized beta variate tends to Φ as the parameters tend to infinity in a fixed ratio (Cramer [5]). It follows that

$$\Pi_n\{n^{\frac{1}{2}}\frac{\theta_1 - rn^{-1}}{[r(n-r)n^{-1}(n+1)^{-1}]^{\frac{1}{2}}} < v\} \to \Phi(v) \tag{15}$$

as $n \to \infty$ on a fixed s sequence, $v \in R$. This, together with condition (i), implies condition (ii). Condition (iii) follows, with a little care, by applying the Lindberg-Levy central limit theorem to $F_{U'|\theta}$ as $m \to \infty$ with $\theta = rn^{-1}$, plus the fact that $\Pi_n(|\theta_1 - rn^{-1}| > \epsilon) \to 0$ as $n \to \infty$ if rn^{-1} is fixed, as it is on a fixed s sequence. Thus (12) holds, and takes the form

$$[\nu_a(r,n) - rn^{-1}]n^2 r^{-\frac{1}{2}}(n-r)^{-\frac{1}{2}}(1-a)^{\frac{1}{2}} \to 2^{-\frac{1}{2}} \tag{16}$$

as $n(1-a) \to \infty$ on a fixed rn^{-1} sequence and $a \to 1$.

Extensive calculations of the dynamic allocation index $\nu_a(r,n)$ have been reported by Gittins and Jones [10]. To carry out such calculations the infinite set of stopping times τ involved in the definition (2) of a dynamic allocation index must be approximated by a finite set. In fact for each n the calculation of $\nu_a(r,n)$ was restricted to values of $\tau \leq N(a)-n$, where $100 \leq N(a) \leq 300$. The effect of this restriction is that the calculations underestimate the value of $\nu_a(r,n)$ to an extent which increases as $n \to N(a)$ and as $a \to 1$. Consequently N(a) was chosen so as to increase with a. These calculations have been used to check the rate at which the left-hand side of (16) approaches its limiting value of 0.707, with the results shown in tables 1 and 2 for $rn^{-1} = 0.5$. In table 1 the values for n > 150 are underestimates because of the approach of a to N(a), and in table 2 the third significant figure is unreliable because of rounding errors.

Table 1. LHS of (16); $rn^{-1} = 0.5$; a = 0.99, 0.95 and 0.9

n	10	25	50	75	100	125	150	175	200	225	250	275
a=0.99 N(a)=300	.339	.441	.506	.538	.558	.570	.579	.585	.588	.583	.565	.514
a=0.95 N(a)=200	.428	.501	.537	.552	.559	.565	.570	.560				
a=0.9 N(a)=150	.428	.481	.506	.512	.519	.518						

Table 2. LHS of (16); $rn^{-1} = 0.5$; $a = 0.8$ and 0.7

n	10	20	40	60	80	90
a=0.8 N(a)=100	.386	.413	.429	.435	.436	.435
a=0.7 N(a)=100	.334	.355	.364	.368	.368	.375

It is interesting to note that, apart from the effect of finite values of N(a), each row of the above tables appears to approach a limit as n increases. This is not predicted by (16), in contrast to the fact that the rate of convergence is less rapid for values of a close to 1. The same pattern is revealed by calculations for other values of rn^{-1}, and it turns out that for suitably chosen functions A_a, B_a and C_a defined on the interval (0,1) the approximation

$$\nu_a(r,n) = rn^{-1} + [A_a(rn^{-1}) + n\,B_a(rn^{-1}) + n^{-1}C_a(rn^{-1})]^{-1} \qquad (17)$$

is very good except for small values of n. Comparing (17) with (16) it is clear that $nr^{-\frac{1}{2}}(n-r)^{-\frac{1}{2}}(1-a)^{\frac{1}{2}}[B_a(rn^{-1})]^{-1}$ should be a good estimate of the limit as $n \to \infty$ of any given row of tables 1 and 2. These estimates are shown in table 3.

Table 3. Limit of LHS of (16) as $n \to \infty$; $rn^{-1} = 0.5$

a	.7	.8	.9	.95	.99
lim(LHS)	.37	.44	.52	.58	.63

These results are certainly consistent with (16), which indicates that the double limit as $a \to 1$ is 0.707. The results for other values of rn^{-1} are very similar, except that the figures for rn^{-1} outside the range (0.05, 0.95) are appreciably lower than those shown in table 3.

Exponential Process

X_i has density $\ell_\theta(x) = \theta e^{-\theta x}$ with respect to Lebesgue measure on R^+. $\Theta = R^+$, $\theta_1 = \theta^{-1}$, $\theta_2 = \theta^{-2}$. Improper prior density π_0 = constant. Π_n has density $\pi_n \propto \theta^n \exp(-\theta n\bar{x}_n)$, which is a gamma density with parameters n and $n\bar{x}_n$, with respect to which θ_1 has mean $\bar{x}_n$ and variance $(n-1)^{-1}\bar{x}_n^2$.

Putting $s_n = \bar{x}_n$, condition (i) again follows from Chebychev's theorem. Condition (ii) follows from the Lindberg-Levy central limit theorem when we note that V' is asymptotically a standardised gamma variate whose distribution function may be expressed as the convolution of n + 1 identical negative exponential distribution functions. U' and θ are independent with respect to P_n, and condition (iii) is thus an immediate consequence of the same theorem. Thus (12) holds and takes the form

$$[\nu_a(\bar{x},n)-\bar{x}]\bar{x}^{-1}\ n(1-a)^{\frac{1}{2}} \to 2^{-\frac{1}{2}} \qquad (18)$$

as $n(1-a) \to \infty$ on a fixed $\bar{x}$ sequence and $a \to 1$. A simple scale invariance argument shows that the left hand side of (18) is actually independent of $\bar{x}(> 0)$.

As yet unpublished calculations of the dynamic allocation index $\nu_a(1,n)$ have been carried out using the dynamic programming iterative procedure described in [6]. This is a different procedure from that used for the Bernoulli process, but truncates the set of stopping times considered in the same way, and with similar effects on the accuracy of the results. As for the Bernoulli process it turns out that the left hand side of (18) tends to a limit as $n \to \infty$ for any fixed value of a. The truncation point N(a) was taken to be 400 for every a, which is sufficient for convergence to the limit to be observed for $a \leq 0.99$. The following limit values were obtained, which are consistent with (18).

Table 4. Limit of LHS of (18) as $n \to \infty$

a	.5	.6	.7	.8	.9	.95	.99
lim(LHS)	.214	.268	.329	.400	.495	.570	.684

6. A THEOREM ON SINGLE MACHINE SCHEDULING

Before presenting our final illustration of the foregoing theory we first prove an apparently new result on single machine scheduling which will be needed.

Single machine scheduling problems may often (see [6] and [7]) be represented by *families of alternative bandit processes* (i.e. generalised discounted multi-armed bandits). The arms represent the jobs to be scheduled and pulling arm i corresponds to allocating the machine to job i. Consider then two discrete-time families of alternative bandit processes A and B differing only in the rewards associated with the various possible changes of state.

There are m bandit processes in each of the two families. The state spaces for each of the bandit processes are quite general, except that each of them includes a state C corresponding to completion of the job. For a bandit process in family A the reward each time it is selected is 0 unless this results in a transition to state C from some other state, in which case the reward is 1. For a bandit process in family B the reward each time it is selected is -1 unless it is in state C, in which case the reward is 0. No transition from state C to any other state is possible. In all cases the discount factor is a.

Thus family A represents the problem A of scheduling m jobs, each with the same terminal reward, so as to maximise the expected total discounted reward. For family B any optimal policy clearly selects a bandit process in state C as soon as one is available. In effect nothing happens after this point: no changes of state and no rewards. Thus family B represents the problem B of scheduling so as to minimise the expected cost incurred up to the first job completion time t_1, assuming that the cost of operating the machine up to time t is $(1-a^t)(1-a)^{-1}$. Equivalently, Ea^{t_1} is maximised.

We shall show that, except in state C, the dynamic allocation indices of the bandit processes in family A are given by a strictly increasing function of their values in family B. Since these indices determine the optimal policies in each case, this implies

Theorem 2. For a discounted single machine scheduling problem in which each job yields the same reward on completion any policy which maximises the expected total reward must also maximise the expected reward from the first job completion.

Proof. For an arbitrary job in state s ($\neq$ C) let $s_0(=s)$, $s_1, s_2, \ldots$ be the sequence of states induced by successive 'pulls'. Let $T = \min\{s_i = C\}$ and $p_i = 1-q_i = P(s_{i+1} = T | s_i)$. The set of stopping times τ considered in the discrete time version of (2) may obviously be restricted for problem A to those for which $\tau \le T$, and for problem B to those for which $P(T<\tau<\infty) = 0$. Moreover there is an obvious 1:1 correspondence between these two restricted sets of stopping times. Let τ_A and τ_B denote corresponding members of the two sets.

Using superscripts to distinguish the R's, W's and ν's associated with problems A and B respectively, it follows from these definitions together with those given in section 2, that

$$R^A_{\tau_A}(s) = E(p_0 + aq_0p_1 + a^2q_0q_1p_2 + \ldots + a^{\tau-1} q_0q_1 \ldots q_{\tau-2}p_{\tau-1}) ,$$

$$W^A_{\tau_A}(s) = -R^B_{\tau_B}(s) = E(1 + aq_0 + a^2q_0q_1 + \ldots + a^{\tau-1} q_0q_1 \ldots q_{\tau-2}) ,$$

and

$$W^B_{\tau_B}(s) = W^A_{\tau_A}(s) + a(1-a)^{-1}R^A_{\tau_A}(s) .$$

where τ is such that $\tau_A = \min\{T,\tau\}$ and $\tau_B = \begin{cases} \tau & \text{if } \tau < T \\ \infty & \text{otherwise} \end{cases}$

$$R^A_{\tau_A}(s) \big/ W^A_{\tau_A}(s) = -a^{-1}(1-a)\left(W^B_{\tau_B}(s) \big/ R^B_{\tau_B}(s) + 1\right) ,$$

and hence, taking the suprema of each side over τ_A and τ_B, that

$$\nu^A_a(s) = -a^{-1}(1-a)[(\nu^B_a(s))^{-1} + 1] .$$

This is the required strictly increasing relationship between ν^A anc ν^B, completing the proof of the theorem.

Corollary 2.1. The expected time up to the first job completion is minimised by using the functions ν^A as priority indices with a = 1.

Proof. Note that if the discount factor a is set equal to 1 problem B becomes that of minimising the expected time up to the first job completion. The corollary follows from the continuity of ν^A in a at a = 1, and the continuity of the expected total reward both in a and over a suitably metrised policy space.

7. A TARGET PROCESS

Problem B of the previous section offers a plausible model of situations in which there is a single objective which may be pursued simultaneously in several different ways, only one of which is requi to succeed. The exponential target process arises in one such situa tion which is common in chemical research, when different formulatio are being tested with the aim of finding one whose activity achieves some target level (Gittins and Jones [9]). If the formulations whic could be tested belong to recognisably different groups, each group corresponds to a possible random sequence of test results defining a bandit process, which in this case will be termed a *target* process

Target processes may be defined in terms of any parametric family of probability distributions. We shall, however, consider on the exponential target process, defined just as the exponential rewa of section 5 except that the only reward which occurs is a reward of 1 when the first of the observations X_i exceeds the target level, which we may define to be 1 by suitable choice of scale. It is a jol as defined in problem A of section 6, if attainment of the target is regarded as completion of the job. As shown by theorem 2 the reward structure is different, but not the nature of an optimal policy, if the aim is to solve problem B.

The expected reward from the next observation when the targe has not yet been reached is $e^{-\theta}$ for a given value of θ. The expectation of $e^{-\theta}$ with respect to the current distribution Π_n with parameters n and $\bar{x}$ is $(n\bar{x})^n(n\bar{x}+1)^{-n}$, which $\to \exp(\bar{x}^{-1})$ as $n \to \infty$. The functions $(ny)^n(ny+1)^{-1}$ and $\exp(y^{-1})$ satisfy the conditions for $G(y,n)$ and $G(y)$ of corollary 1.1, as a little manipulation shows, and conditions (i) to (iii) were checked in section 5. However, the situation differs from that considered in the discussion leading to the limit (14) because the sequence of expected rewards $(n\bar{x})^n(n\bar{x}+1)^{-n}$ terminates as as soon as the target is reached.

In fact (14) may be applied directly to a modified target process for which rewards occur for every observation X_i which exceeds the target, rather than just for the first such observation. In the notation of section 6 the dynamic allocation index for a target process modified in this way and in state s may be written

$$\nu_a(s) = \sup_{\tau>0} \frac{E(p_0+ap_1+a^2p_2+\ldots+a^{\tau-1}p_{\tau-1})}{E(1+a+a^2+\ldots+a^{\tau-1})} \tag{19}$$

The expression for $\nu^A(s)$ may be obtained by writing $a^r q_0 q_1 \ldots q_{r-1}$ where a^r occurs in (19), $(r=1,2,\ldots,\tau-1)$.

This observation suggests a method of obtaining an approximation for $\nu_a^A(\bar{x},n)$. Provided q_r only varies slowly with r, which is true for large values of n, we should find that

$$\nu_a^A(\bar{x},n) \simeq \nu_{aq}(\bar{x},n)$$

where $q = q_0 = 1 - (n\bar{x})^n(n\bar{x}+1)^n = q(\bar{x},n)$. Finally then, plugging all this into (14), our conjecture is that

$$[\nu_a^A(\bar{x},n) - 1 + q(\bar{x},n)]\; n\bar{x}\, \exp(\bar{x}^{-1})[1-aq(\bar{x},n)]^{\frac{1}{2}} \to 2^{-\frac{1}{2}} \tag{20}$$

as $n(1-aq) \to \infty$ on a fixed $\bar{x}$ sequence and $aq \to 1$. It remains a conjecture because we have not shown that it is legitimate to replace q_r by q_0 when aq is close to 1, even for large values of n.

Using the dynamic programming iterative procedure described in [6], Jones [12] has made some calculations of the function $\nu_a^A(\bar{x},n)$ when $a = 1$. The calculations actually produce values of $\bar{x}$ for which $\nu_a^A(\bar{x},n)$ takes a given value ν as n varies. Again the set of stopping times τ is truncated by restricting τ to values $\leq N-n$ for some fixed N. These results have been used to provide values of the left hand side of (20), with the modification that the factor $n^{\frac{1}{2}}\bar{x}\exp(\bar{x}^{-1})$ was replaced by $[(n\bar{x})^n(n\bar{x}+2)^{-n} - (n\bar{x})^{2n}(n\bar{x}+1)^{-2n}]^{-\frac{1}{2}}$. The ratio of these two factors $\to 1$ as $n \to \infty$, the latter being the reciprocal of the variance with respect to Π_n of $e^{-\theta}$, the expected reward given θ. The results of these calculations are shown in table 5.

Table 5. Modified LHS of (20); ν = 0.001 and 0.1

n	ν=0.1 N=500	n	ν=0.001 N=1000	n	ν=0.001 N=1000
1	.0224	1	.001	500	.407
2	.0356	2	.003	550	.408
5	.0564	5	.010	600	.410
10	.0712	10	.034	650	.405
20	.0826	20	.088	700	.401
50	.0918	50	.189	750	.391
100	.0960	100	.266	800	.374
150	.0980	150	.307	850	.348
200	.0976	200	.340	900	.308
250	.0981	250	.362	950	.235
300	.0997	300	.375	980	.145
350	.0996	350	.383	995	.049
400	.0986	400	.394	998	.009
450	.0988	450	.404		
480	.0980				
495	.0758				
498	.0343				

The striking and unpredicted feature of this table is the evidence for a limit as $n \to \infty$ for fixed ν, except when n is near enough to N for the truncation of τ values to have an effect. This is directl related to the observed convergence of the left hand side of (18) to a limit as $n \to \infty$ and a is fixed. The convergence of these limits to 0.707 as $\nu \to 0$, i.e. as $q \to 1$, is clearly very slow, if it occurs at all.

REFERENCES

1. J.A. BATHER (1962) Bayes procedures for deciding the sign of a normal mean. *Proc. Camb. Phil. Soc.* **58**, 599-620

2. J.A. BATHER (1983) Optimal stopping of Brownian motion: a comparison technique. *H. Chernoff 60'th Birthday Festschrift*, edited by D. Siegmund et al., Academic Press, New York

3. H. CHERNOFF (1961) Sequential tests for the mean of a normal distribution. *Proceedings of the fourth Berkeley Symposium on Statistics* **1**, 79-91

4. H. CHERNOFF (1968) Optimal stochastic control. *Sankhya Series* **30**, 221-252

5. H. CRAMER (1946) *Mathematical Methods of Statistics*, Princeton University Press, Princeton

6. J.C. GITTINS (1979) Bandit processes and dynamic allocation indices. *J. Roy. Statist. Soc. B* **41**, 148-177

7. J.C. GITTINS (1982) Forwards induction and dynamic allocation indices. *Deterministic and Stochastic Scheduling*, (editors M.A.H. Dempster et al.) 125-156, Reidel, Dordrecht, Holland

8. J.C. GITTINS and D.M. JONES (1974) A dynamic allocation index for the sequential design of experiments. *Progress in Statistics* (editor J. Gani), 241-266, North Holland, Amsterdam

9. J.C. GITTINS and D.M. JONES (1974) *A Dynamic Allocation Index for New-Product Chemical Research.* Technical Report-Mgt. Stud. 13, Cambridge University Engineering Department

10. J.G. GITTINS and D.M. JONES (1979) A dynamic allocation index for the discounted multi-armed bandit problem. *Biometrika* 66, 561-5

11. K.D. GLAZEBROOK (1982) On a sufficient condition for superprocesses due to Whittle. *J. Appl. Prob.* **19**, 99-110

12. D.M. JONES (1974) *Search Procedures for Industiral Chemical Research.* Ph.D. thesis, Cambridge University

13. F.P. KELLY (1981) Multi-armed bandits with discount factor near one - the Bernoulli case. *Annals of Statist.* **9**, 987-1001

14. L. LE CAM (1953) On some asymptotic properties of maximum likelihood estimates and related Bayes' estimates. *Univ. Calif. Publ. Statist.* **1**, 277-329

15. D.V. LINDLEY (1960) Dynamic programming and decision theory. *Appl. Statist.* **10**, 39-51

16. D.V. LINDLEY (1965) *Introduction to Probability and Statistics from a Bayesian Viewpoint, Part 2. Inference*, Cambridge University Press, Cambridge

17. P. NASH (1973) *Optimal Allocation of Resources between Research Projects.* Ph.D. thesis, Cambridge University

18. P. WHITTLE (1980) Multi-armed bandits and the Gittins index. *J. Roy. Statist. Soc. B* **42**, 143-149

19. P. WHITTLE (1981) Arm-acquiring bandits. *Ann. Prob.* **9**, 284-292

THE ROLE OF DYNAMIC ALLOCATION INDICES IN THE EVALUATION OF SUBOPTIMAL STRATEGIES FOR FAMILIES OF BANDIT PROCESSES

K.D. Glazebrook

Department of Statistics, University of Newcastle upon Tyne

Newcastle upon Tyne, U.K.

1. Introduction

A family of N alternative bandit processes $\{(\Omega_j, P_j, C_j, \alpha);\ j=1,2,\ldots N\}$ is a cost-discounted Markov decision process with the following special features:

(a) Its state at time $t\in\mathbb{N}$ is $x(t)=\{x_1(t),\ x_2(t),\ldots,x_N(t)\}$ where $x_j(t)\in\Omega_j$, $1\leq j\leq N$. State space Ω_j may be finite, countable or continuous.

(b) The action space A is $\{a_1,a_2,\ldots,a_N\}$. Action a_j denotes the choice of bandit process j. An action is taken at each time $t\in\mathbb{N}$.

(c) If action a_j is taken at time $t\in\mathbb{N}$ only the jth component of $x(t)$ changes. Hence $x_i(t+1)=x_i(t)$, $i\neq j$, and $x_j(t+1)$ is determined according to a probabilistic law of motion $P_j\{x_j(t)\}$.

(d) The transition of the process under action a_j described in (c) incurs a cost $\alpha^t C_j\{x_j(t),\ x_j(t+1)\}$ where $0\leq\alpha\leq 1$. The costs are assumed to be bounded.

(e) An optimal strategy is a rule for choosing actions which minimises the total expected cost.

Such decision processes model situations where a decision maker has a choice among N options (here referred to as bandit processes) which evolve stochastically in time. Fields of application include research planning (see for example Nash [14]), stochastic job-shop scheduling (Gittins and Glazebrook [4]), search problems (Gittins [3]), computer scheduling (Bruno and Hofri [2]) and the sequential design of experiments (Gittins [3]). Under the last heading we include many of the classical bandit problems.

Gittins and Jones (1974) demonstrated that optimal strategies for families of alternative bandit processes are determined by a collection of dynamic allocation indices (DAI's) γ_j, $1\leq j\leq N$, as follows:

THEOREM 1.1 Given a family of alternative bandit processes $\{(\Omega_j, P_j, C_j, \alpha);\ j=1,2,\ldots,N\}$ there exist real-valued functions γ_j on Ω_j with the property that action a_i is optimal when the process is in state $x(t)$ if and only if

$$\gamma_i\{x_i(t)\}=\min_j \gamma_j\{x_j(t)\}$$

A characterisation of index γ_j is given by Gittins and Glazebrook [4] as follows: Consider the decision process $\{(\Omega_j, P_j, C_j, \alpha);\ j=1,2,\ldots,N\}$ under the strategy which always chooses action a_j then

$$\gamma_j(x_j)=\inf_{\tau>0}\left(E\left[\sum_{t=0}^{\tau-1}\alpha^t C_j\{x_j(t),\ x_j(t+1)\}\right]\{1-E(\alpha^\tau)\}^{-1}\right), \qquad (1)$$

all expectations being conditional on $x_j(0)=x_j$ and the infimum being over all positive stopping times. Gittins and Jones [5] originally characterised index $\gamma_j(x_j)$ as an equivalent stopping cost for bandit process j when in state x_j. Nash [14] proved that the infimum in (1) is attained by stopping time $\tau_j^*(x_j)$ where

$$\tau_j^*(x_j)=\inf_{t>0}\left[t;\ \gamma_j\{x_j(t)\}>\gamma_j(x_j)\right] \qquad (2)$$

The optimal strategies determined by DAI's according to Theorem 1.1 have been criticised by, for example, Bather [1] on the grounds that they are often difficult to construct and apply. It is certainly true that in many problems the calculation of the indices in (1) is highly non-trivial. It is also true that in some problem areas optimality criteria other than that of minimised discounted costs may be of interest. For both of these reasons the problem of the evaluation of suboptimal strategies for families of alternative bandit processes is an important one. Section 2 contains a discussion of this problem. It turns out that DAI's play an important and intuitive rôle.

In Sections 3 and 4 we discuss the application of the material in Section 2 to two important problems in stochastic scheduling. This work also serves to illustrate how the theory copes with problems in continuous time (Section 3) and problems involving processes more complex than families of alternative bandit processes (Section 4).

2. The evaluation of suboptimal strategies

Here we obtain a bound on the loss incurred when implementing an arbitrary stationary strategy for a family of bandit processes instead of an optimal strategy based on DAI's.

The main result here is Theorem 2.1. Let S be a stationary strategy for the family of N alternative bandit processes $\{(\Omega_j, P_j, C_j, \alpha); j=1,2,\ldots,N\}$ and let $\gamma\{S, x(t)\}$ be the DAI of the bandit process chosen by strategy S when the family of processes is in state x(t). If strategy S is non-deterministic then $\gamma\{S, x(t)\}$ is a random variable. We denote by S* an optimal strategy for the family of processes (i.e. one determined according to the statement of Theorem 1.1), by C(S*), C(S) the total expected costs incurred by strategies S* and S respectively and by $C\{S;[\zeta,\eta)\}$ the contribution to C(S) from applying strategy S throughout the (possibly) random time interval $[\zeta,\eta)$. Theorem 2.1 yields an upper bound on the extent to which strategy S is suboptimal.

THEOREM 2.1

$$C(S)-C(S^*)\le E_S\left(\sum_{t=0}^{\infty}\alpha^t[\gamma\{S,x(t)\}-\min_j\gamma_j\{x_j(t)\}]\right), \qquad (3)$$

E_S denoting an expectation over all realisations of the process $\{(\Omega_j, P_j, C_j, \alpha);\ j=1,2,\ldots,N\}$ under strategy S.

Proof

Denote by S_j a strategy which at time t=0 takes action a_j and which from time t=1 onwards chooses actions according to a DAI strategy of the kind described in Theorem 1.1. We shall first obtain an upper bound on $C(S_j) - C(S^*)$ for the case when

$$\gamma_j\{x_j(0)\}>\min_i\gamma_i\{x_i(0)\} \tag{4}$$

To that end, we define $\tilde{S}_j$ to be a strategy which at times $t=0,1,\ldots,\tau^*_j\{x_j(0)\}-1$ takes action a_j, where as in (2) we have that

$$\tau^*_j\{x_j(0)\}=\inf_{t>0}\left[t;\gamma_j\{x_j(t)\}>\gamma_j\{x_j(0)\}\right],$$

and which from time $t=\tau^*_j\{x_j(0)\}$ chooses actions according to a DAI strategy. We firstly observe that

$$C(\tilde{S}_j)\geq C(S_j) \tag{5}$$

and proceed to find an upper bound on $C(\tilde{S}_j)-C(S^*)$.

For the process under strategy $\tilde{S}_j$ we define stopping time $\tau^*_j\{x_j(0)\}+\hat{\tau}_j$ by

$$\tau^*_j\{x_j(0)\}+\hat{\tau}_j = \inf_{t>\tau^*_j\{x_j(0)\}}\left[t;\ \min_{i\neq j}\gamma_i\{x_i(t)\}>\gamma_j\{x_j(0)\}\right].$$

We write

$$\begin{aligned}C(\tilde{S}_j) &= C\{\tilde{S}_j;[0,\tau^*_j\{x_j(0)\})\} + C\{\tilde{S}_j;[\tau^*_j\{x_j(0)\},\ \tau^*_j\{x_j(0)\}+\hat{\tau}_j)\}\\ &\quad + C\{\tilde{S}_j;[\tau^*_j\{x_j(0)\}+\hat{\tau}_j,\ \infty)\}.\end{aligned} \tag{6}$$

It is not difficult to show that, as a consequence of the definition of strategy $\tilde{S}_j$ and the characterisation of DAI's given in (1) that

$$C\{\tilde{S}_j;\ [0,\tau^*_j\{x_j(0)\})\} = \gamma_j\{x_j(0)\}\{1-E(\alpha^{\tau^*_j\{x_j(0)\}})\} \text{ and} \tag{7}$$

$$\begin{aligned}\gamma_j\{x_j(0)\}[E(\alpha^{\tau^*_j\{x_j(0)\}})-E(\alpha^{\tau^*_j\{x_j(0)\}+\hat{\tau}_j})] &\geq C\{\tilde{S}_j;[\tau^*_j\{x_j(0)\},\ \tau^*_j\{x_j(0)\}+\hat{\tau}_j)\}\\ &\geq \min_i\ \gamma_i\{x_i(0)\}[E(\alpha^{\tau^*_j\{x_j(0)\}}) - E(\alpha^{\tau^*_j\{x_j(0)\}+\hat{\tau}_j})]\end{aligned} \tag{8}$$

The following comparison between strategies $\tilde{S}_j$ and DAI strategy S* is based on the fact that two successive parts of the process under $\tilde{S}_j$ (i.e. the choices of action during $[0,\ \tau^*_j\{x_j(0)\})$ and $[\tau^*_j\{x_j(0)\},\ \tau^*_j\{x_j(0)\}+\hat{\tau}_j))$ occur in the reverse order under S* with no change to what happens to the process at other times. Hence

$$\begin{aligned}C(S^*) &= [E(\alpha^{\tau^*_j\{x_j(0)\}})]^{-1}C\{\tilde{S}_j;[\tau^*_j\{x_j(0)\},\ \tau^*_j\{x_j(0)\}+\hat{\tau}_j)\}\\ &\quad + E(\alpha^{\hat{\tau}_j})C\{\tilde{S}_j;[0,\tau^*_j\{x_j(0)\})\} + C\{\tilde{S}_j;\ [\tau^*_j\{x_j(0)\}+\hat{\tau}_j,\infty)\}\end{aligned} \tag{9}$$

It is an easy consequence of (6) - (9) that

$$C(\tilde{S}_j) - C(S^*) \leq \gamma_j\{x_j(0)\} - \min_i\ \gamma_i\{x_i(0)\}$$

and hence from (5) that

$$C(S_j) - C(S^*) \leq \gamma_j\{x_j(0)\} - \min_i\ \gamma_i\{x_i(0)\} \tag{10}$$

Note that inequality (10) holds trivially when

$$\gamma_j\{x_j(0)\} = \min_i\ \gamma_i\{x_i(0)\} \tag{11}$$

since then strategies S_j and S* coincide. Since (4) and (11) exhaust all possibilities, inequality (10) always holds and gives the desired upper bound on $C(S_j)-C(S^*)$.

Inequality (3) now follows from (10) by standard conditioning and summation arguments.

In general the expression on the right hand side of inequality (3) may be thought of as a measure of the extent to which strategy S departs from a DAI strategy, this measure being related to the difference between the DAI's of the bandit processes chosen by S and those which (under S) have minimal indices at each stage.

Several examples of the use of Theorem 2.1 now follow.

EXAMPLE 1 Many of the standard procedures for the calculation of DAI's are approximative and/or iterative. See for example those described by Gittins [3] and Gittins and Jones [6]. Some of these procedures have guarantees on accuracy. We may decide, using one of these procedures, to calculate DAI's to within Δ of their true values. A question of interest to us will be how much we lose by basing a strategy on these approximate values instead of exact ones. A trivial application of Theorem 2.1 serves to show that this loss cannot exceed $2\Delta(1-\alpha)^{-1}$.

EXAMPLE 2 This example concerns rules for sampling one at a time from $N(\geq 2)$ Bernoulli populations, population i having unknown probability of success p_i.

If we assume that the p_i's have independent Beta priors and that success (failure) on the tth trial yields a reward of α^{t-1} (zero) then the problem of finding the Bayes sequential decision rule may be formulated as a family of N alternative bandit processes and Theorem 1.1 applies. Hence for this problem optimal strategies are determined by a collection of DAI's.

Robbins [15], followed by Bather [1] proposed the study of strategies which are asymptotically optimal. Such strategies guarantee that the observed proportion of successes converges to $\max_i p_i$ as the number of trials becomes infinite. In general, DAI strategies are not asymptotically optimal.

Glazebrook [8] demonstrated the existence of rules based on randomised DAI's which are both asymptotically optimal and ε-Bayes for any $\varepsilon>0$. The analysis which yields that conclusion involves the evaluation of a suboptimal randomised strategy. The direct approach to this evaluation used by Glazebrook [8] is rather cumbersome. The main result is now seen to be a straightforward consequence of Theorem 2.1.

EXAMPLE 3 A class of strategies with some intuitive appeal are the one-step look ahead rules, sometimes called myopic strategies. Such strategies always seek to minimise the expected cost incurred in the next time instant with no regard for what might happen in the more distant future. Glazebrook [10] has utilised Theorem 2.1 to evaluate a one-step look ahead rule for a simple one-armed bandit problem with discounted costs. Not surprisingly the performance of the rule deteriorates as the discount rate α increases to 1.

3. A Bayesian model in single machine scheduling

To demonstrate how the theory extends to continuous time cases we now consider an important single machine scheduling problem, discussed previously by Gittins and Glazebrook [4] and Glazebrook [11].

A job shop consists of a single machine and a set of N jobs to be processed on it. The processing times for the jobs $\{P_1, P_2, \ldots, P_N\}$ are independent random variables with finite expectations, P_j having absolutely continuous density $\theta_j f_j(\theta_j t)$ characterised by unknown scale parameter θ_j. The distribution function of P_j is $F_j(\theta_j t)$ and θ_j has a prior density which is uniform on $[0, M_j]$. No rewards are earned or costs incurred except at job completions. If job j is completed at time t the reward is $R_j e^{-\beta t}$ where $\beta > 0$.

At each time $t \geq 0$ the machine processes jobs according to one of the set of admissible controls at time t,

$$\Omega(t) = \{(\pi_1, \pi_2, \ldots, \pi_N);\ \pi_i = 0,\ i \in J(t) \text{ and } \sum_{i=1}^{N} \pi_i = 1\}$$

where J(t) is the set of jobs which have been completed at time t. Suppose that strategy S chooses vector-valued control $\pi(t)$ at time t and that $x_j(t)$ is the amount of processing which job j has received prior to t, then $\pi_j(t) = x_j'(t)$.

The objective of processing is to earn the largest expected total reward possible, the expectation being taken over all realisations of the process and with respect to the prior densities.

An appropriate continuous time version of Theorem 1.1 serves to demonstrate that an optimal strategy is such that $\pi_j(t) > 0$ if and only if $j \notin J(t)$ and

$$\gamma_j\{x_j(t), M_j\} = \min_i \gamma_i\{x_i(t), M_i\} \qquad (12)$$

the minimisation in (12) being over all $i \notin J(t)$ and the DAI $\gamma_i(x_i, M_i)$ being given by

$$\gamma_i(x_i, M_i) = \inf_{s>0} \left[-R_i \left[\int_{r=0}^{s} \int_{\theta_i=0}^{M_i} e^{-\beta r} \theta_i f_i\{\theta_i(x_i+r)\} d\theta_i\, dr \right] \left(\int_{r=0}^{s} \int_{\theta_i=0}^{M_i} e^{-\beta r} [1 - F_i\{\theta_i(x_i+r)\}] d\theta_i\, dr \right)^{-1} \right]$$

Hence, under an optimal processing strategy at time t the machine will be shared between those of the uncompleted jobs which have a minimal value of the index $\gamma_i\{x_i(t), M_i\}$.

In the analysis which follows we shall see that for all $x_i > 0$

$$\gamma_i(x_i, M_i) \to -R_i x_i^{-1}, \quad M_i \to \infty$$

This strongly suggests that any strategy, $\hat{S}$ say, which processes at time t uncompleted jobs with the largest associated value of $R_i\{x_i(t)\}^{-1}$ should perform well when all the M_i's are large. This is substantiated in Theorem 3.2. Note, too, the intuitive appeal of such a strategy, favouring as it does jobs with high returns (i.e. large R_i) and small amounts of past processing (i.e. small $x_i(t)$). Indeed $\hat{S}$ provides an interesting variant of the celebrated Smith's rule [16].

In order to evaluate strategy $\hat{S}$ we need a continuous time version of Theorem 2.1. Theorem 3.1 may be deduced from Theorem 2.1 by considering a sequence of appropriately chosen discrete time problems, allowing the discrete-time quantum to tend to zero.

<u>THEOREM 3.1</u> If strategy $\hat{S}$ chooses control $\hat{\pi}(t)$ at time t and if strategy S* denotes an optimal strategy then

$$C(\hat{S}) - C(S^*) \leq E_{\hat{S}} \left\{ \int_0^\infty \beta^{-1} e^{-\beta t} \sum_{j=1}^{N} [\gamma_j\{x_j(t), M_j\} - \min_i \gamma_i\{x_i(t), M_i\}]\, \hat{\pi}_j(t) dt \right\}, \quad (13)$$

the minimisation in (13) being over all $i \notin J(t)$.

In order to explore the implications of Theorem 3.1 for our scheduling problem we need to introduce functions ε_i given by

$$\varepsilon_i(t) = t\{1-F_i(t)\}\left[\int_0^t \{1-F_i(u)\}\,du\right]^{-1}, \quad t>0, \quad 1\le i\le N.$$

Note that it follows from the assumption that P_i has finite mean that

$$\varepsilon_i(t)\to 0, \quad t\to\infty.$$

Theorem 3.2 asserts that strategy $\hat{S}$ is optimal in an asymptotic sense.

THEOREM 3.2 If $x_i(0)>0$, $1\le i\le N$, then

$$C(\hat{S}) - C(S^*) \le E_{\hat{S}}\left\{\int_{t=0}^{\infty} \beta^{-1} e^{-\beta t} \sum_{i=1}^{N} R_i\{x_i(t)\}^{-1} \varepsilon_i[M_i\{x_i(t)\}^{-1}]\,dt\right\}$$

$$\to 0, \quad \min_i M_i \to \infty .$$

Proof

We define the completion rate of job i evaluated at $x_i>0$ to be

$$\rho_i(x_i, M_i) = \int_{\theta_i=0}^{M_i} \theta_i f_i(\theta_i x_i)\,d\theta_i \left[\int_{\theta_i=0}^{M_i}\{1-F_i(\theta_i x_i)\}\,d\theta_i\right]^{-1}$$

Simple analysis serves to show that for $x_i>0$

$$x_i^{-1} \ge \rho_i(x_i, M_i) = x_i^{-1}\{1-\varepsilon_i(M_i x_i^{-1})\}. \tag{14}$$

From (14) it can be deduced, using basic properties of the DAI's concerned, that

$$-R_i x_i^{-1} \le \gamma_i(x_i, M_i) \le -R_i x_i^{-1}\{1-\varepsilon_i(M_i x_i^{-1})\}. \tag{15}$$

It now follows from (15) and the definition of strategy of $\hat{S}$ that for all $t\ge 0$

$$\sum_{j=1}^{N} [\gamma_j\{x_j(t), M_j\} - \min_i \gamma_i\{x_i(t), M_i\}]\hat{\pi}_j(t)$$

$$\le \sum_{j=1}^{N} R_j\{x_j(t)\}^{-1}\varepsilon_j[M_j\{x_j(t)\}^{-1}] \tag{16}$$

Theorem 3.2 is a simple consequence of Theorem 3.1 and (16)

4. Single machine scheduling with precedence constraints

Our aim in this example is to demonstrate the application of Theorem 2.1 (or, more properly, an extension of it) to a process which is more complex than a family of alternative bandit processes. This process models a problem of allocating a single machine to a collection of stochastic jobs in discrete time when there are technological or other constraints which govern the order in which jobs may be processed. Work on this model and related ones has been done by Bruno and Hofri [2], Glazebrook [7] and Meilijson and Weiss [13].

A collection of stochastic jobs is available for processing by a single machine. A precedence relation R (i.e. a partial ordering on the job set) whose digraph representation is an out-forest delimits the set of jobs which can be processed at any time. We write $(a,b)\in R$ if we require that job a be completed before job b can be processed. Job a is said to be an immediate predecessor of job b if $(a,b)\in R$ and if there exists no job c for which $(a,c)\in R$ and $(c,b)\in R$. The requirement that the digraph representation of R be an out-forest is equivalent to the requirement that each job should have at most one immediate predecessor. It is not difficult to show that an out-forest may be decomposed into a collection of m connected components called out-trees each of which has exactly one source (i.e. a job with no predecessors). In what follows $\{i1, i2, \ldots, in_i\}$ is the set of jobs corresponding to the vertices in component i, job i1 being that one which has no predecessors at time $t=0$, $1 \le i \le m$.

We formulate the scheduling problem as a Markov decision process as follows:

(a) The state of the process at time $t \in \mathbb{N}$ is $x(t) = \{x_1(t), x_2(t), \ldots, x_m(t)\}$ where $x_i(t) \in \{(-1) \cup \mathbb{N}\}^{n_i}$ is the state of component i at time t. Further, we have that

$$x_i(t) = \{x_{i1}(t),\ x_{i2}(t), \ldots, x_{in_i}(t)\}$$

where $x_{ij}(t) \in \{(-1) \cup \mathbb{N}\}$ is the state of job ij at time t. In fact

$$x_{ij}(t) = \begin{cases} -1, & \text{if job ij has been completed by time t;} \\ & \text{the number of units of processing which job ij} \\ & \text{has received by time t otherwise, } 1 \le j \le n_i,\ 1 \le i \le m. \end{cases}$$

Note that $x_{ij}(0)=0$, $1 \le j \le n_i$, $1 \le i \le m$.

(b) Action a_{ij} is the allocation of the machine to job ij and is admissible at time $t \in \mathbb{N}$ if and only if

(i) $(ik, ij) \in R \Rightarrow x_{ik}(t) = -1$ and

(ii) $x_{ij}(t) \ne -1$.

Hence a job is admissible for processing at time t if it has not been completed by then and if all of its predecessors have been completed.

(c) If job ij is processed at time $t \in \mathbb{N}$ the probability of its completion during $[t, t+1)$ is given by

$$\rho_{ij}\{x_{ij}(t)\} \triangleq P\{x_{ij}(t+1) = -1 \mid x_{ij}(t) \ne -1\} = 1 - P\{x_{ij}(t+1) = x_{ij}(t)+1 \mid x_{ij}(t) \ne -1\}$$

where ρ_{ij} is the completion rate for job ij.

(d) Should job ij be completed during time $[t, t+1)$ there is a reward (i.e. a negative cost) $\alpha^{t+1} R_{ij}$.

(e) An optimal strategy is any rule for choosing admissible actions which maximises the total expected reward.

The following result, a refined version of one due to Glazebrook [7], asserts the existence of a DAI-like optimal strategy for the above problem. The proof uses Theorem 1.1 together with a simple induction on the number of jobs.

<u>THEOREM 4.1</u> There exist real-valued functions γ_{ij} an $\{(-1) \cup \mathbb{N}\}$ with the property that action a_{ij} is optimal for decision process (a)-(e) at time t if and only if

(i) a_{ij} is admissible at time t and

(ii) if $a_{k\ell}$ is also admissible at time t then $\gamma_{k\ell}\{x_{k\ell}(t)\} \ge \gamma_{ij}\{x_{ij}(t)\}$

A general discussion of when DAI-like strategies are optimal for processes which are more complex than families of alternative bandit processes may be found in Whittle [17] and Glazebrook [9].

We now proceed to describe how suitable DAI's may be calculated for our scheduling problem. Denote by $J(ij)$ the subset of $\{i1, i2, \ldots, in_i\}$ consisting of job ij together with all the jobs which ij precedes. Denote by $S(ij)$ any admissible stationary strategy for the jobs in $J(ij)$ (viewed in isolation from all other jobs). When strategy $S(ij)$ is applied to $J(ij)$ then the resultant process is a bandit process which has a DAI $\gamma_{ij}\{x_{ij}, S(ij)\}$, calculated according to (1), when job ij is in state x_{ij}. The DAI γ_{ij} in Theorem 4.1 is given by

$$\gamma_{ij}(x_{ij}) = \inf_{S(ij)} \gamma_{ij}\{x_{ij}, S(ij)\}, \tag{17}$$

the infimum being taken over all possible choices of admissible stationary strategy $S(ij)$. The calculation of $\gamma_{ij}(x_{ij})$ is typically very complicated, involving as it does infima over both stopping times and strategies. Hence our main concern, namely the evaluation of suboptimal strategies for this problem, is an important one.

Theorem 4.2, concerning suboptimal strategies, extends Theorem 2.1 to this more complex decision process. Its proof will not be given. The assumption of out-forest precedence constraints simplifies matters to such an extent that the proof is in all essentials the same as that for Theorem 2.1. Again $\gamma\{S, x(t)\}$ denotes the DAI of the job which strategy S chooses when the process is in state $x(t)$, DAI's being calculated according to (17).

<u>THEOREM 4.2</u> If S is a stationary admissible strategy for decision process (a) - (e) then

$$C(S) - C(S^*) \le E_S\Big(\sum_{t=0}^{\infty} \alpha^t[\gamma\{S, x(t)\} - \min_{ij} \gamma_{ij}\{x_{ij}(t)\}]\Big), \tag{18}$$

the minimisation in (18) being taken over all ij for which a_{ij} is admissible at time t.

The most obvious way of utilising Theorem 4.2 as an effective evaluation tool is to produce bounds on $\gamma_{ij}(x_{ij})$. This is the subject of Lemma 4.3. In this result we use $\hat{\gamma}_{ij}(x_{ij})$ to denote the DAI of job ij in state x_{ij} when job ij is viewed in isolation from all the other jobs. Alternatively, $\hat{\gamma}_{ij}(x_{ij})$ may be viewed as the DAI appropriate for job ij in the problem identical to ours in all respects except that $R=\emptyset$. The important point is that $\hat{\gamma}_{ij}(x_{ij})$ is much simpler to calculate than $\gamma_{ij}(x_{ij})$ - indeed we have that for $x_{ij} \neq -1$,

$$\hat{\gamma}_{ij}(x_{ij}) = \inf_{s>0} \Big(-R_j(1-\alpha)^{-1} \sum_{t=1}^{s} \alpha^t p_{ij}(x_{ij}+t) \prod_{r=0}^{t-1}\{1-p_{ij}(x_{ij}+r)\}\Big[\sum_{t=0}^{s-1} \alpha^t \prod_{r=0}^{t}\{1-p_{ij}(x_{ij}+r)\}\Big]^{-1}\Big)$$

where the infimum is over the positive integers.

<u>LEMMA 4.3</u> (a) If $J(ij) \neq (ij)$ then

$$\hat{\gamma}_{ij}(x_{ij}) \ge \gamma_{ij}(x_{ij}) \ge \min\{\hat{\gamma}_{ij}(x_{ij}), \min_{ik} \hat{\gamma}_{ik}(0)\} \tag{19}$$

where the minimisation inside $\{\ \}$ is over all $ik \in J(ij)\setminus(ij)$.

(b) if $J(ij)=ij$ then $\hat{\gamma}_{ij}(x_{ij}) = \gamma_{ij}(x_{ij})$.

<u>Proof</u>

(a) We first consider the left-hand inequality in (19). From (1) and (17) we may write

$$\gamma_{ij}(x_{ij}) = \inf_{S(ij)} \inf_{\tau>0} \Delta\{x_{ij}, S(ij), \tau\} \tag{20}$$

for suitably chosen quantity Δ. It is not difficult to see from (1) that

$$\hat{\gamma}_{ij}(x_{ij}) = \inf_{S(ij)} \inf_{T\geq\tau>0} \Delta\{x_{ij}, S(ij), \tau\} \tag{21}$$

where T is the random time at which job ij is completed under application of admissible strategy S(ij). Inspection of (20) and (21) plainly reveals that $\hat{\gamma}_{ij}(x_{ij}) \geq \gamma_{ij}(x_{ij})$.

In order to prove the right-hand inequality, consider a scheduling problem identical to the one described in (a) - (e) except that $R = \emptyset$. Denote by $\mathcal{S}(ij)$ the set of admissible stationary strategies for the jobs in J(ij) for this new problem. It is not difficult to show that

$$\min\{\hat{\gamma}_{ij}(x_{ij}), \min_{ik}\hat{\gamma}_{ik}(0)\} = \inf_{S} \inf_{\tau>0} \Delta\{x_{ij}, S, \tau\} \tag{22}$$

where the infimum is over all $S \in \mathcal{S}(ij)$ and the minimisation in $\{\ \}$ is over all $ik \in J(ij)\backslash(ij)$.

A comparison of the right-hand sides of (20) and (22) shows that the first infimum in (22) is over a larger class of strategies than is the case in (20). The right-hand inequality in (19) then follows.

(b) is trivial.

We now consider the problem of evaluating strategy $\hat{S}$ which processes jobs according to the simple indices $\hat{\gamma}_{ij}$, i.e. $\hat{S}$ chooses a_{ij} at time + if

(i) a_{ij} is admissible at time t, and

(ii) if $a_{k\ell}$ is also admissible at time t then $\hat{\gamma}_{k\ell}\{x_{k\ell}(t)\} \geq \hat{\gamma}_{ij}\{x_{ij}(t)\}$

We think of $\hat{S}$ as a quasi-myopic strategy for this problem in that it chooses among the currently admissible jobs on the basis of an evaluation of those jobs alone. Hence it does not consider possible future rewards accruing from jobs not currently admissible.

In order to simplify our evaluation of $\hat{S}$ on the basis of Theorem 4.2 and Lemma 4.3 we now assume that all jobs within a single component are identical, i.e. that

$$ik \in J(ij) \Rightarrow \rho_{ik} \equiv \rho_{ij},\ R_{ik} = R_{ij}. \tag{23}$$

It follows from (23) that

$$\hat{\gamma}_{ik} \equiv \hat{\gamma}_{ij}. \tag{24}$$

Now define

$$\hat{\gamma}^{+}_{ij}(x_{ij}) = \begin{cases} \max\{\hat{\gamma}_{ij}(x_{ij}) - \hat{\gamma}_{ij}(0),\ 0\}, & x_{ij} \in \mathbb{N}; \\ 0, & x_{ij} = -1. \end{cases}$$

It follows from Lemma 4.3 and from (24) that

$$\hat{\gamma}_{ij}\{x_{ij}(t)\} - \gamma_{ij}\{x_{ij}(t)\} \leq \hat{\gamma}^{+}_{ij}\{x_{ij}(t)\}$$

provided that $x_{ij}(t) \neq -1$. Corollary 4.4 now follows from Theorem 4.2.

COROLLARY 4.4

$$C(\hat{S}) - C(S^*) \le E_{\hat{S}}\left[\sum_{t=0}^{\infty} \sum_{i=1}^{m} \sum_{j=1}^{n_i} \alpha^t \hat{\gamma}_{ij}^{+}\{x_{ij}(t)\}\right]$$

A more detailed discussion of the simplifying features of out-forest precedence constraints may be found in Glazebrook [12]. Please note in conclusion that Theorem 4.2 is a special result relating to this stochastic scheduling problem. Work continues on the general problem of evaluating suboptimal strategies for complex decision processes for which an index-type strategy is optimal.

References

[1] J.A. BATHER, Randomized allocation of treatments in sequential experiments, J.R. Stat. Soc. Ser. B, 43, 265-292, 1981.

[2] J. BRUNO and M. HOFRI, On scheduling chains of jobs on one processor with limited preemption, SIAM J. Comput., 4, 478-490, 1975.

[3] J.C. GITTINS, Bandit processes and dynamic allocation indices, J.R. Stat. Soc. Ser. B, 41, 148-177, 1979.

[4] J.C. GITTINS and K.D. GLAZEBROOK, On Bayesian models in stochastic scheduling, J. Appl. Probab., 14, 556-565, 1977.

[5] J.C. GITTINS and D.M. JONES, A dynamic allocation index for the sequential design of experiments, Progress in Statistics, ed. J. Gani, North-Holland, Amsterdam, 1974.

[6] J.C. GITTINS and D.M. JONES, A dynamic allocation index for the discounted multi-armed bandit problem, Biometrika, 66, 561-565, 1979.

[7] K.D. GLAZEBROOK, Stochastic scheduling with order constraints, Int. J. Syst. Sci., 7, 657-666, 1976.

[8] K.D. GLAZEBROOK, On randomized dynamic allocation indices for the sequential design of experiments, J.R. Stat. Soc. Ser. B, 42, 342-346, 1980.

[9] K.D. GLAZEBROOK, On a sufficient condition for superprocesses due to Whittle, J. Appl. Probab., 19, 99-110, 1982.

[10] K.D. GLAZEBROOK, On the evaluation of suboptimal strategies for families of alternative bandit processes, J. Appl. Probab. (to appear).

[11] K.D. GLAZEBROOK, Myopic strategies for Bayesian models in stochastic scheduling (submitted).

[12] K.D. GLAZEBROOK, On the evaluation of stochastic scheduling problems with order constraints, Int. J. Syst. Sci. (to appear).

[13] I. MEILIJSON and G. WEISS, Multiple feedback at a single server station, Stochastic Processes Appl. 5, 195-205, 1977.

[14] P. NASH, Optimal allocation of resources between research projects, Ph.D. thesis, Cambridge University, 1973.

[15] H. ROBBINS, Some aspects of the sequential design of experiments, Bull. Am. Math. Soc., 58, 527-535, 1952.

[16] W.E. SMITH, Various optimisers for single-stage production, Nav. Res. Logist. Q. 3, 59-66, 1956.

[17] P. WHITTLE, Multi-armed bandits and the Gittins index, J.R. Stat. Soc. Ser. B, 42, 143-149, 1980.

ON THE DISCRETIZATION TECHNIQUE FOR OPTIMAL DISCOUNTED CONTROL OF THE WIENER PROCESS

K. Helmes

Institut für Angewandte Mathematik
Universität Bonn

1. INTRODUCTION

As has been emphasized by G.N. Saridis in his lecture (see this volume) stochastic control problems constitute sub-tasks to be solved in complex intelligent control systems. In this note we shall present a discretization technique for solving stochastic control models. The method yields explicit solutions for well structured lower dimensional problems. The power of the technique is demonstrated by looking at a specific control problem which has also been analyzed using different techniques (cf.[1],[2],[3] and [9]).

We consider a continuous time inventory model where the state $x(t) \in \mathbb{R}$, the inventory level of a commodity at time $t \in [o,\infty)$, evolves according to the stochastic differential equation

(1) $$dx_t = u_t(x)dt + dw_t \ , \ t \geq o \ , \ x(o) = x \ ,$$

where (u_t) denotes any measurable non-anticipative control bounded in absolute value by 1. We treat the backlog case, thus allowing negative inventory levels. The set of all admissible strategies will be denoted by $\mathcal{U}$. The noise (w_t) is assumed to be a standard Brownian motion defined on a probability space $(\Omega,\mathcal{F},P)$ adapted to an increasing family of sub σ-algebras $\mathcal{F}_t$ of $\mathcal{F}$. For each initial value $x \in \mathbb{R}$ and control $u \in \mathcal{U}$ there is a unique (weak) solution to equation (1). The value of the strategy u at time t can be thought of as the ordering rate of the commodity. There is a cost $\phi(x_t)$ for being in the 'wrong' state x_t, where we assume ϕ to satisfy the following conditions:

(A_1) ϕ is an even, non-negative and monotone increasing function (on $\mathbb{R}^+$) which satisfies a polynomial growth condition, i.e.

(2) $$|\phi(x)| = O(|x|^m)$$

for some $m > o$.

There is also the cost $|u_t|$ for using the control u_t. For the infinite horizon problem both costs are discounted by the factor $\exp(-\alpha t), \alpha > 0$. The controller has to choose an admissible law (u_t) so as to minimize the expected discounted total cost over $[0,\infty)$

$$(3) \qquad J(x;u) = E_x[\int_0^\infty \exp(-\alpha t)[\phi(x_t) + |u_t(x)|]dt].$$

In [9] it has been verified using the Bellman equation for this problem together with a heuristic principle of smooth fit that the 'physically obvious' control, which is to push with full force (to the direction of zero) if the inventory level is outside a certain neighbourhood of the origin while to exert no control at all if x_t is in this neighbourhood,

$$(4) \qquad u^*(t,x) = \begin{cases} -1 & x_t > b \\ 0 & |x_t| \leq b, \\ +1 & x_t < -b, \end{cases}$$

is indeed the optimal one. The cutoff point b separating the active region from the dead zone has been characterized in terms of the function ϕ through a transcendental equation, see (13) below. In order to prove existence and uniqueness of a solution to this transcendental equation the following conditions, more stringent than (A_1), had to be imposed on ϕ:

(A_1') ϕ is an even, non-negative, $C^2(\mathbb{R})$ function whose second derivative is decreasing with distance from the origin and which is uniformly convex in the sense that

$$0 < \kappa_1 \leq \phi(x) \leq \kappa_2 \ , \ x \in \mathbb{R} \ ,$$

for some positive constants κ_1, κ_2.

The idea which underlies the discretization technique is to approximate the continuous time problem P by a specific sequence of discrete time inventory models (P_h). Then for each discrete time problem the fundamental equation of dynamic programming yields a fixed point equation for the corresponding value function V_h from which qualitative properties of V_h, like symmetry and monotonicity, are derived once it is shown that the non-linear "one-step-look-ahead operator" is a contraction. The properties of V_h imply that for each problem P_h there exists an optimal policy u_h which is a monotone decreasing function on $\mathbb{R}^+$, asymmetric around zero and which has the form

$$(5)\qquad u_h^*(x) = \begin{cases} -1 & x > b_h + 1/h, \\ \text{linear in between,} & \\ o & |x| \leq b_h \quad , \\ \text{linear in between,} & \\ +1 & x < -b_h - 1/h, \end{cases}$$

Thus, at least for a subsequence, the functions (u_h^*) converge pointwise to a function u^* which has form (2), the case $b = \infty$ still being possible. By a general limit theorem on the approximation of continuous time stochastic control problems by controlled Markov chains optimality of u^* follows. Based on further analysis of the discrete time problems we can, under additional conditions on ϕ, viz.

(A_2) ϕ is an even, non-negative, convex function on $\mathbb{R}$ which satisfies a polynomial growth condition

fully describe the value function of the continuous time problem and verify existence of an optimal 'bang-bang' control. A necessary and sufficient condition, based on comparison between the given problem and the one where no control is applied, is given which rules out the simple case of no switching, i.e. $b = \infty$. If there is a switching point b it is a solution to the transcendental equation mentioned above.

In this note proofs of almost all of these statements will only be sketched; for details of the proofs we refer to [7] or [8]. General methods for approximating stochastic control models by (discrete time) controlled Markov chains were developped, among others, by Kushner [10]. The specific discretization technique used here has also been applied in [4] and [5] to analyze other classes of stochastic control problems.

2. THE DISCRETE TIME MODELS

Fix $h > 0$ and consider the discrete time system with dynamic equation

$$(6)\qquad x_{i+1} = x_i + u_i h + \varepsilon_i \ , \quad i = 0,1,2,\ldots,$$

and initial condition $x_o = x$. Here the (ε_i) are independent *normal* random variables with mean value zero and variance h, which are defined on (Ω,F,P). As controls (u_i) we admit all 'nonanticipating' random variables on (Ω,F,P) with values in $[-1,+1]$, i.e. all random variables u_i which are independent of the future disturbances $\varepsilon_{i+1}, \varepsilon_{i+2}, \ldots$. Let Σ_h denote the set of all admissible strategies $\pi = (u_o, u_1, \ldots)$. When a policy $\pi \in \Sigma_h$ is chosen E_x^π denotes expectation with respect to the probability measure determined by π and the transition law described by (6). The objective is to choose a strategy π in such a way as to minimize

$$(7) \qquad J(x;h,\pi) := E_x^\pi \left[\sum_{k=o}^{\infty} h \exp(-\alpha hk) \left[\phi(x_k) + |u_k| \right] \right],$$

i.e. to find a strategy π with average cost equal to the value of the problem

$$(8) \qquad V_h(x) := \inf_{u\pi \in \Sigma_h} \left\{ J(x;h,\pi) \right\}, \quad x \in \mathbb{R}.$$

Let $K_h(x)$ denote the Gaussian density with mean value zero and variance h and let the $*$-operator denote convolution of two functions. By the general theory of dynamic programming, f.i. [6], p.150, V_h satisfies the functional equation

$$(9) \qquad V_h(x) = \inf_{u \in [-1,1]} \left\{ \exp(-\alpha h) V_h * K_h(x+uh) + h\phi(x) + h|u| \right\}.$$

The crucial feature of the approximation scheme is the normal distribution of the disturbances (ε_i). Choosing normally distributed errors is more than just a natural discretization of the Wiener process; in fact, the selection made is dictated by the necessity of getting functions V_h having 'nice' properties. Specifically, this way it is possible to prove convergence of Markov chains to diffusions with non-continuous drift terms (see Theorem 1, below) and to show that some of the properties which ϕ has are inherited by the functions V_h. The proof of this fact is based on equation (9) and on the contraction principle well known in the case of a bounded reward function. Sinse ϕ is unbounded this proof has to be slightly modified. Instead of the space of bounded measurable functions we consider the space B_h of those measurable functions for which

$$\|f\| := \sup_{x \in \mathbb{R}} \frac{|f(x)|}{1+R_h(x)} < \infty .$$

Here

$$R_h(x) = E_x \left[\sum_{k=o}^{\infty} \exp(-\alpha kh/2) h\phi(\bar{x}_k) \right],$$

the $\alpha/2$ - resolvent of the process $(\bar{x}_k)$,

$$\bar{x}_{k+1} = \bar{x}_k + h\,\mathrm{sign}(\bar{x}_k) + \varepsilon_k\ , \ \bar{x}(o) = x\ , \quad k=0,1,2,\ldots,$$

applied to $h\phi$.

Lemma 1. *The non-linear operator T_h defined by*

$$(10)\quad T_h f(x) = h\phi(x) + \inf_{-1\le u\le 1} \left\{ h|u| + \exp(-\alpha h) K_h * f(x+uh) \right\}$$

is a monotone contraction operator on the Banach space $(B_h, \|\cdot\|)$.

Proof. The proof runs along the lines proving this fact for a bounded reward function. The only extra work to be done consists in checking the estimate

$$e^{-\alpha h} R_h \phi * K_h(x+hu(x)) \le e^{-\alpha h/2} R_h \phi(x)$$

for any measurable function $u: \mathbb{R} \to [-1,+1]$. But this estimate is implied by the very definition of $(\bar{x}_k)$ and the properties of $R_h \phi * K_h$.

An application of Lemma 1 and Banach's fixed point theorem is the next result.

Lemma 2. *If ϕ satisfies A_1, A_2 resp., then V_h fulfills A_1, A_2 respectively.*

An immediate consequence of Lemma 2, the functional equation (9) and monotonicity (on $\mathbb{R}^+$) of $V_h * K_h$ is the assertion:

Lemma 3. *If ϕ satisfies A_1 then for each problem P_h there exists a unique (stationary) optimal strategy $u_h^*(x)$ given by minimizing (pointwise) the expression*

$$\exp(-\alpha h) V_h * K_h(x+uh) + h|u|,\ x \in \mathbb{R}$$

over $u \in [-1,+1]$. The function $u_h^*(x)$ is asymmetric around zero, monotone decreasing on $\mathbb{R}^+$ and has the form, $b_h \in [o,\infty]$,

$$u_h^*(x) = \begin{cases} o & x \in [o, b_h), \\ \text{linear} & x \in [b_h, b_h + 1/h], \\ -1 & x \in [b_h + 1/h, \infty), \\ \text{asymmetric around zero.} & \end{cases}$$

The switching point $b_h \in [o,\infty]$ is determined by the equation

$$(11)\qquad \dot{V}_h * K_h(b_h) = \exp(\alpha h)$$

If ϕ satisfies condition (A_1) existence of an optimal (for problem P) control u^* having at most two switching points can be proved.

Theorem 1. Let ϕ satisfy A_1. For the continuous time problem (1), (3) there exists an optimal control u^* given by Eq. (4). The switching point $b\in[0,\infty]$ is an accumulation point of the sequence (b_h) described in Lemma 3.

Proof. By Lemma 3, the 'sequence' of functions (u_h^*) contains a subsequence which has u^*, u^* given by (4), as its limit function. Since u^* is discontinuous at most at two points, $\pm b$ ($b=\infty$ is possible), the Markov chains (a subsequence) corresponding to (u_h^*) converge to a diffusion with drift term u^* and variance 1, see [4]; for a more general convergence result of this kind see [7]. Moreover, the values $V_h(x)$ converge to $V(x)$. This follows from weak convergence of the probability measures associated with the Markov chains and a straightforward truncation argument applied to ϕ.

3. FORMULA FOR THE VALUE FUNCTION

In addition to A_2 we shall assume from now on that Φ is sufficiently smooth, i.e. at least C^2. Then, by Lemma 2, each function V_h satisfies A_2 and thus, by Theorem 1, the value function of the continuous time problem

$$V(x) = \inf_{u\in\mathcal{U}}\left[E_x \int_0^\infty \exp(-\alpha t)[\phi(x_t) + |u_t(x)|]dt\right],$$

$x(t)$ defined by (1), possesses these properties as well; in particular, $V(x)$ is a continuous function. Differentiability of V will be verified next.

Lemma 4. Let $(b_{h'})$ be a convergent subsequence of the switching points (b_h) with limit value $b<\infty$. Then the value function V is differentiable and

$$\dot{V}(b) = 1 . \tag{12}$$

Proof. Let (W_k), $(\bar{W}_k)$ resp., denote a simple random walk with Gaussian transition probabilities having mean value zero and variance h, mean value -h respectively. It follows from Eq. (9), Lemma 3 and the properties of the functions - (Y_k) denotes (W_k) or $(\bar{W}_k)$ -

$$x \longrightarrow E_x\left[\sum_{k=0}^{\infty} h \exp(-\alpha hk)\phi(Y_k)\right]$$

that the functions $(\dot{V}_h)$ are unifromly Lipschitzian on any compact interval $I \subset \mathbb{R}$. By the Arzela-Ascoli theorem $\{\dot{V}_h\}$ is compact in the sense of uniform convergence on I. So take a uniformly convergent subsequence (again denoted by $\dot{V}_h$) converging to a limit $\tilde{V}$. Taking the limit in the equality

$$V_h(x) - V_h(o) = \int_o^x \dot{V}_h(y)\,dy \quad , \; x > o,$$

we get

$$V(x) - V(o) = \int_o^x \tilde{V}(y)\,dy \; .$$

Therefore, $\tilde{V} \equiv \dot{V}$, $(\dot{V}_h)$ has a single limit point and $\dot{V}_h \to \dot{V}$ uniformly on I. Since

$$\dot{V}_h(b_h) = h\;\dot{\phi}(x) + h|u_h^*(x)| + e^{-\alpha h}\; K_h * \dot{V}_h(x+hu_h(x))$$

(cf. Fleming, Rishel: Deterministic and Stochastic Optimal Control, p. 53) Eq. (11) (see Lemma 3) yields Eq. (12).

Now we are able to give a complete description of the value function V; for the proof see [7] or [8]. Let J(x;u) denote the cost incurred when controlling the continuous time system according to the rule u. Put $j(x) := J(x;o)$, $p(x) := J(x;+1)$, $q(x) := J(x;-1)$ and

$$I := \sup_{x\in\mathbb{R}} |j(x)|.$$

Theorem 2. *Let ϕ be a smooth function which satisfies A_2.*
(i) If $I \leq 1$ then $u^* \equiv 0$ is an optimal control and the value function is given by

$$V(x) = \int_{-\infty}^{\infty} \phi(y)\, \upsilon_\alpha(x,y)\,dy,$$

$\upsilon_\alpha(x,y)$ the α-resolvent density of standard Brownian motion, i.e.

$$\upsilon_\alpha(x,y) = (2\alpha)^{-1/2}\, exp\left[-\sqrt{2\alpha}\,|x-y|\right].$$

(ii) If $I > 1$ then u^* as specified by (4) is an optimal control; the switching point b is a solution to the transcendental equation

$$\text{(13)} \qquad tanh(b\sqrt{2\alpha}) = \frac{j(b) - 1}{\sqrt{2\alpha}\left[p(b) - q(b) + \beta^{-1}(1-\dot{q}(b))\right]}$$

where $\beta = \sqrt{1+2\alpha} - 1$. The value function is given by

$$V(x) = \begin{cases} q(x) - \frac{1-q(b)}{\beta} \exp[-\beta(x-b)], & x \geq b \\ j(x) - \frac{j(b) - 1}{\sqrt{2\alpha}\sinh(b\sqrt{2\alpha})} \cosh[x\sqrt{2\alpha}], & 0 \leq x \leq b \\ V(-x), & x \leq 0 \end{cases}$$

REFERENCES

[1] BATHER, J.A.: *A continuous time invertory model*, J. Appl.Prob., vol. 3, pp. 538-549, 1966.

[2] BATHER, J.A.: *A diffusion model for the control of a dam*, J. Appl.Prob., vol. 5, pp. 55-71, 1968.

[3] BENEŠ, V.E., SHEPP, L.A.and WITSENHAUSEN, H.S.: *Some solvable stochastic control problems*, Stochastics, vol. 4, pp. 39-83, 1980.

[4] CHRISTOPEIT, N. and HELMES, K.: *On Beneš' bang-bang control problem*, to appear in Appl. Math. Optimization, 1982.

[5] CHRISTOPEIT, N. and HELMES, K.: *Optimal control for a class of partially observable systems*, to appear in Stochastics, 1982.

[6] DYNKIN, E.B. and YUSHKEVICH, A.A.: *Controlled Markov Processes*, Springer-Verlag, New York, 1979.

[7] HELMES, K.: *Approximation gesteuerter Diffusionsprozesse durch stochastische dynamische Programme*, Habilitationsschrift, Bonn, 1982.

[8] HELMES, K.: *Optimal discounted control for a continuous time inventory model*, submitted for publication to JOTA, 1982.

[9] KARATZAS, I.: *Optimal discounted linear control of the Wiener process*, JOTA, vol. 31, pp. 431-440, 1980.

[10] KUSHNER, H.J.: *Probability methods for approximations in stochastic control and for elliptic equations*, Academic Press, New York, 1977.

Dedicated to Professor Octav Onicescu
on the occasion of his 90th birthday

ASYMPTOTIC PROPERTIES OF LEARNING MODELS

by

Marius Iosifescu
Centre of Mathematical Statistics, Bucharest, Romania

1. Introduction

It is already well known that all stochastic models for learning studied up to now enter the following general scheme (cf. Norman (1972)). The behaviour of the subject on trial $n = 0,1,\ldots$ is determined by his state S_n (an indicator of the subject's response tendencies) at the beginning of the trial. S_n is a random variable taking on values in a measurable state space $(S,\underline{S})$. On trial n an event E_{n+1} occurs that results in a change of state. E_{n+1} is a random variable taking on values in a measurable event space $(E,\underline{E})$ and specifies those occurrences on trial n that affect subsequent behaviour. Typically, E_{n+1} includes a specification of the subject's response and its observable outcome or payoff. To represent the fact that the occurrence of an event effects a change of state we consider a measurable mapping v of $S \times E$ into S and postulate that $S_{n+1} = v(S_n,E_{n+1})$, $n \geq 0$. Finally, we assume that the conditional probability distribution of E_{n+1} given $E_n, S_n, \ldots$ depends only on the state S_n and denote it $Q(S_n,\cdot)$.

The collection $((S,\underline{S}),(E,\underline{E}), v,Q)$ is said to be a (general stochastic) learning model (Norman (1972), p.24). To obtain various special models we should just particularize $S,E,v,$ and Q.

A general learning model is nothing but a (homogeneous) random system with complete connections (RSCC), a concept set up by Romanian authors (see, e.g., Iosifescu (1963), Iosifescu and Theodorescu (1969), Norman (1972)) who also contributed substantially to its study. Historically, the first example of an RSCC is represented by Onicescu and Mihoc's chains with complete connections.

The aim of this paper is to present the asymptotic results which have been obtained for RSCC's (i.e., for the general stochastic learning model) during the last 10 years. A detailed treatment will be found in the forthcoming book by Grigorescu and Iosifescu (1983).

It is not difficult to prove in a formal manner the existence of the two sequences of random variables $(E_n)_{n \geq 1}$ and $(S_n)_{n \geq 0}$ associated with $((S,\underline{S}),(E,\underline{E}),v,Q)$. To be precise, for any $s \in S$ there exists a probability space $(\Omega,\underline{K}, P_s)$ on which the two sequences are defined (with $S_0=s$), and

$$P_s(E_1 \in A) = Q(s,A),$$

$$P_s(E_{n+1} \in A \mid S_n, E_n,\ldots,S_1,E_1) = Q(S_n,A), \quad P_s\text{-a.s.}$$

for any $A \in \underline{E}$ and $n \geq 1$. The sequence $(E_n)_{n \geq 1}$ is therefore an infinite order chain while, as easily seen, the sequence $(S_n)_{n \geq 0}$ is a homogeneous Markov chain.

An interesting feature is that the Markov chain associated with an RSCC is, except for certain degenerated cases, of a special type that cannot be brought under established theories for Markov chains (see Subsection 2.1.).

We close these introductory considerations by noticing that there is a somewhat abusive usage of the nomenclature "learning" in contexts different from that of psychology. In our opinion a precise and a natural criterion for labeling a situation as a learning one should be <u>the possibility of describing it in the framework of random systems with complete connections.</u>

It should also be stressed that "learning" becomes more and more a subject of general interest. This is witnessed, e.g. , by the recent report to the Club of Rome by Botkin, Elmandjra, and Malitza (1980).

2. Two invariance principles

2.1. Denote by $(E^n,\underline{E}^n)$ the n-fold product of the measurable space $(E,\underline{E})$ and for

$$e^{(n)} = (e^{(n-1)}, e_n) = (e_1,\ldots,e_n) \in E^n$$

define $v(s,e^{(n)})$ recursively as $v(v(s,e^{(n-1)}),e_n)$, $n \geq 2$.

For any $A \in \underline{E}^n$ put

$$Q_n(s,A) = P_s((E_1,\ldots,E_n) \in A).$$

Assume (cf.Norman (1972), pp.31 and 66-67) that

(a) S is a bounded separable metric space with metric d and $\underline{S}$ is the class of Borel subsets of S ;

(b) $d(v(s',e), v(s'',e)) \leq d(s',s'')$ for all $e \in E$, $s',s'' \in S$;

(c) there exists a non-negative number C such that

$$|Q(s',A) - Q(s'',A)| \leq Cd(s',s'')$$

for all $A \in \underline{E}$, $s',s'' \in S$;

(d) there exist a probability p on $\underline{E}$ and an $a > 0$ such that $Q(s,A) \geq ap(A)$ for all $A \in \underline{E}$, $s \in S$;

(e) there is a $j \geq 1$ such that

$$\sup_{s' \neq s''} \int_E p(de_1) \ldots \int_E p(de_j) \frac{d(v(s',e^{(j)}), v(s'',e^{(j)}))}{d(s',s'')} < 1$$

Conditions (a) to (e) above ensure the existence of a probability P_∞ on $\underline{K}$ such that

$$\lim_{n \to \infty} Q_{m+n}(s, E^n \times A) = P_\infty((E_1, \ldots, E_m) \in A)$$

uniformly with respect to $m \geq 1$, $A \in \underline{E}^m$. In fact the rate of convergence in the above equation is exponential. The sequence $(E_n)_{n \geq 1}$ is φ-mixing under both P_∞ and P_s whatever $s \in S$. (Under P_∞ it is a strictly stationary one.)

As to the sequence $(S_n)_{n \geq 0}$ it is a so-called Doeblin-Fortet (Markov) chain. This means that the transition operator U defined as usual on $B(S,\underline{S})$ — the Banach space of real valued bounded and $\underline{S}$-measurable functions on S under the supremum norm — by the equation[1]

$$(Uf)(s) = M_s(f(S_1)) = \int_E Q(s,de)f(v(s,e)), \quad f \in B(S,\underline{S}),$$

has the following two properties :

I. U maps $L(S)$ into itself boundedly with respect to $\|\cdot\|$. Here $L(S)$ denotes the Banach algebra of real valued bounded Lipschitz functions f on S under the norm $\|f\| = m(f) + |f|$ with

$$m(f) = \sup_{s' \neq s''} \left| \frac{f(s') - f(s'')}{d(s',s'')} \right| ,$$

$$|f| = \sup_{s \in S} |f(s)| .$$

II. There are a natural integer k and non-negative numbers $r < 1$ and R such that for all $f \in L(S)$

1) M_s denotes the mean value operator with respect to P_s.

$$m(U^k f) \leq r\, m(f) + R\,|f| .$$

2.2. Let F be a real valued measurable function on $E \times E \times \ldots$ and put $f_n = F(E_n, E_{n+1}, \ldots)$, $n \geq 1$. Assume without any loss of generality that [2] $M_\infty f_1 = 0$. (Otherwise we consider the function $F - M_\infty f_1$ instead of F.) Put $T_0 = 0$, $T_n = f_1 + \ldots + f_n$, $n \geq 1$, and define the random functions

$$X_n^C : X_n^C(t) = \frac{1}{\sigma\sqrt{n}} \left(T_{[nt]} + (nt - [nt])\, f_{[nt]+1}\right), \quad t \in [0,1] ,$$

$$X_n^D : X_n^D(t) = \frac{1}{\sigma\sqrt{n}} T_{[nt]}, \quad t \in [0,1] , \quad n \geq 1 ,$$

where σ is a positive number to be specified in the theorem below.

<u>Theorem 1</u> (Popescu (1977)). Assume that $M_\infty f_1^2 < \infty$ and

$$\sum_{n \geq 1} M_\infty^{1/2} \left[f_1 - M_\infty (f_1 \mid \sigma((E_i)_{1 \leq i \leq n})) \right]^2 < \infty ,$$

where $\sigma(E_i)_{1 \leq i \leq n}$ denotes the σ-algebra generated by the random variables E_i, $1 \leq i \leq n$. Then the series

$$\sigma^2 = M_\infty f_1^2 + 2 \sum_{n \geq 1} M_\infty f_1 f_{n+1}$$

converges absolutely and $\sigma \geq 0$. If $\sigma > 0$ then both random functions X_n^C and X_n^D converge in distribution under P_s to the standard Brownian motion process for any $s \in S$.

(Note that convergence in distribution of X_n^C and X_n^D under P_∞ has been proved by Billingsley (1968), p.184.)

Theorem 1 can be used as in Billingsley (1968), pp.77-83 to prove limit theorems for various functionals of the partial sums T_n. For details see Popescu (1977).

Assume now that F is a real valued measurable function on E^k for some given natural integer k and put $f_n = F(E_n, \ldots, E_{n+k-1})$, $n \geq 1$. Assume also that $M_\infty f_1 = 0$.

<u>Theorem 2</u> (Iosifescu (1972), Heyde and Scott (1973)). Assume that $M_\infty f_1^2 < \infty$. Then the series

$$\sigma^2 = M_\infty f_1^2 + 2 \sum_{n \geq 1} M_\infty f_1 f_{n+1}$$

converges absolutely and $\sigma \geq 0$. If $\sigma > 0$ the sequence

$$\left(X_n^C / \sqrt{2 \log \log n}\right)_{n \geq 3}$$

---

2) M_∞ denotes the mean value operator with respect to P_∞.

considered as a subset of $C[0,1]$ is precompact and its derived set coincides with K (= the set of absolutely continuous functions $x \in C[0,1]$ such that $x(0) = 0$ and $\int_0^1 [x'(t)]^2\, dt \leq 1$) P_s-almost surely for any $s \in S$.

Obviously, Theorem 2 contains as a special case the classical law of the iterated logarithm for the sequence $(f_n)_{n\geq 1}$.

2.3. Consider now an RSCC satisfying conditions (a) to (e) for which S is a compact metric space. Under these conditions (cf. Norman (1972), p.40) there exists a probability q on $\underline{S}$ such that

$$P_\infty(A) = \int_S q(ds)\, P_s(A),\ A \in \underline{K}.$$

In fact q is the (unique) stationary distribution of the Markov chain $(S_n)_{n\geq 0}$.

Let $G \in C(S)$—the space of continuous functions on S—and put $g_n = G(S_n)$, $n \geq 1$. Assume without any loss of generality that $M_\infty g_1 = 0$. Put $U_0 = 0$, $U_n = g_1 + \dots + g_n$, $n \geq 1$, and define the random functions

$$Y_n^C : Y_n^C(t) = \frac{1}{\tau\sqrt{n}}\left(U_{[nt]} + (nt - [nt])g_{[nt]+1}\right),\ t \in [0,1],$$

$$Y_n^D : Y_n^D(t) = \frac{1}{\tau\sqrt{n}}\, U_{[nt]},\ t \in [0,1],\quad n \geq 1,$$

where τ is a positive number to be specified in the theorem below.

<u>Theorem 3.</u> The limit

$$\lim_{n\to\infty} \frac{1}{n} M_\infty U_n^2 = \tau^2$$

exists for any $G \in C(S)$. (In the case where $G \in L(S)$ the value of τ^2 is $M_\infty g_1^2 + 2\sum_{n\geq 1} M_\infty g_1 g_{n+1}$.) If $\tau > 0$ then both random functions Y_n^C and Y_n^D converge in distribution under P_∞ to the standard Brownian motion process. The sequence

$$\left(Y_n^C / \sqrt{2\log\log n}\right)_{n\geq 3}$$

considered as a subset of $C[0,1]$ is precompact and its derived set coincides with K defined above P_∞-almost surely.

Theorem 3 is a consequence of results of Heyde and Scott (1973) and Scott (1971). Note that for $G \in L(S)$ Popescu (1976) proved convergence in distribution of Y_n^C and Y_n^D under P_s for any $s \in S$.

It should be mentioned that the probability q can be expressed in closed form only in a few special cases. See Marliss and McGregor (1971) and the references therein for the case of the two experimenter-controlled events Bush-Mosteller learning model. With a few exceptions the (limiting) stationary distributions associated with different values of the parameters of the model are purely singular. This does not appear to be very surprising in the light of a recent result of Zamfirescu (1981) according to which most (= all except those in a set of first Baire category) continuous functions of uniformly bounded variation have an almost everywhere vanishing derivative .

References

Billingsley,P.(1968) Convergence of Probability Measures. New York: Wiley.

Botkin,J.W., Elmandjra, M. and Malitza,M. (1980) No Limits to Learning. Bridging the Human Gap: A Report to the Club of Rome. Oxford: Pergamon Press.

Grigorescu,Ş. and Iosifescu,M.(1983) Dependence with Complete Connections and Its Applications. Bucharest : Scientific and Encyclopaedic Publishing House.(In Romanian)

Heyde,C.C. and Scott,D.J.(1973) Invariance principles for the law of the iterated logarithm for martingales and processes with stationary increments. Ann.Probab. 1, 428-436 .

Iosifescu,M.(1963) Random systems with complete connections with an arbitrary set of states. Rev.Math.Pures Appl. 8, 611-645; Addenda ibid. 9(1964), 91-92.(In Russian)

Iosifescu,M.(1972) On Strassen's version of the loglog law for some classes of dependent random variables. Z.Wahrsch.Verw.Gebiete 24, 155-158.

Iosifescu,M. and Theodorescu,R.(1969) Random Processes and Learning. Berlin: Springer.

Marliss,G.S. and McGregor,J.R.(1971) The construction of limiting distributions of response probabilities. J.Appl.Probab. 8, 757-766.

Norman,M.F.(1972) Markov Processes and Learning Models. New York: Academic Press.

Popescu,Gh.(1976) A functional central limit theorem for a class of Markov chains. Rev.Roumaine Math.Pures Appl. 21, 737-750.

Popescu,Gh.(1977) Functional limit theorems for random systems with complete connections. Proc.Fifth Conf.Probability Theory (Braşov, 1974), pp.261-275. Bucharest: Publishing House of the

Romanian Academy.

Scott,D.J.(1971) Central limit theorems for martingales and for processes with stationary increments, using a Skorokhod representation approach. Adv.in Appl.Probab. 5, 119-137.

Zamfirescu,T.(1981) Most monotone functions are singular. Amer.Math. Monthly 88, 47-49.

On the infinitesimal characterization of monotone stopping problems in continuous time

A. Irle
Institut für Mathematische Statistik
Einsteinstraße 62
4400 Münster
West Germany

1. Introduction and the Markov case

Consider the problem of optimal stopping for a real-valued stochastic process $(X_t)_{t\in T}$ with a.s. right-continuous paths defined on a complete probability space (Ω,A,P) and adapted to a right-continuous filtration $(A_t)_{t\in T}$. We consider the continuous time case, i.e. $T = [0,\infty)$.

Let us first assume $X_t = f(Z_t,t)$, where $(Z_t)_{t\in T}$ (with respect to $(A_t)_{t\in T}$) is a Markov process with a.s. right-continuous paths taking values in some topological state space E and f is a measurable real-valued function on $E\times T$, such that the following limit

$$\lim_{h\downarrow 0} (E(f(Z_{t+h}, t+h) \mid Z_t=z) - f(z,t))/h$$

exists for all (z,t). Denoting this limit by $Af(z,t)$ we obtain a real-valued function Af on $E\times T$. If f belongs to the domain of the weak infinitesimal operator of the space-time process then the following holds:

$$(1.1) \qquad (f(Z_t,t) - f(Z_0,0) - \int_0^t Af(Z_s,s)ds)_{t\in T} \text{ is a martingale with respect to } (A_t)_{t\in T};$$

see e.g. Athreya and Kurtz [1]. Let us remark that (1.1) can sometimes be shown for f not in the domain of the weak infinitesimal operator, see Rosenkrantz [4], and furthermore, that (1.1) and the optional sampling theorem imply Dynkin's well-known identity for any stopping time τ with $E\tau<\infty$ (see e.g. Athreya and Kurtz [1]):

$$E\, f(Z_\tau,\tau) = E(\int_0^\tau Af(Z_s,s)ds + f(Z_0,0))$$

f, belonging to the domain of the weak infinitesimal operator, is bounded by definition, but Dynkin's identity may also be extended to certain unbounded f, see Shapiro and Wardrop [6]. This paper contains the definition of "monotone case" in the problem of optimal stopping for $(f(Z_t,t))_{t\in T}$, but let us make some remarks, before stating this definition:

Shapiro and Wardrop consider problems of minimization which leads to the reverse inequality in the definition of the set B_t, since we consider problems of maximization.

Relations $\subset$, $=$, resp. $\leq$, $=$ between measurable sets, resp. random variables will be considered to hold a.s.. In the following we will refer to some results holding under suitable technical conditions. This will include that certain exceptional sets of probability zero can be chosen independent of the time parameter.

Let us now set for $t \in T$

$$B_t = \{Af(Z_t,t) \leq 0\}.$$

Then the problem of optimal stopping is in the monotone case if the following holds:

(1.2) $$B_t \subset B_{t+h} \text{ for all } t, h \in T \text{ and } \bigcup_{t \in T} B_t = \Omega.$$

In the monotone case it makes sense to define the infinitesimal look-ahead rule,

$$\rho = \inf \{t \in T: Af(Z_t,t) \leq 0\} = \inf \{t \in T: \omega \in B_t\}$$

and, using Dynkin's identity, the next result is immediate (see Shapiro and Wardrop [6], also Ross [5]).

In the monotone case the following holds under suitable technical conditions:

If Dynkin's identity holds for ρ, then $Ef(Z_\rho,\rho) \geq Ef(Z_\tau,\tau)$ for all stopping times τ for which Dynkin's identity is valid.

A different condition of monotonicity for stopping problems - called weakly monotone case - is introduced in Irle [3] for general (i.e. no assumptions on the distribution) stochastic processes with continuous time parameter.
Defining $C_t = \{X_t \geq \operatorname{ess\,sup}_{s > t} E(X_s \mid A_t)\}$, the stopping problem is in the weakly monotone case, if the following holds:

(1.3) $$C_t \subset C_{t+h} \text{ for all } t, h \in T \text{ and } \bigcup_{t \in T} C_t = \Omega.$$

Then we may define

$$\sigma = \inf \{t \in T: X_t \geq \operatorname{ess\,sup}_{s > t} E(X_s \mid A_t)\} = \inf \{t \in T: \omega \in C_t\}$$

and it is shown in Irle [3], that under suitable technical conditions in the weakly monotone case $EX_\sigma \geq EX_\tau$ holds for all stopping times τ with $\liminf_{t \to \infty} \int_{\{\tau > t\}} X_t^- \, dP = 0$.

We will next show (in the Markov situation) that the monotone case is in fact a special case of the weakly monotone case.

(1.4) Proposition: Assume that (1.1) and (1.2) hold. Then $B_t = C_t$ for all $t \in T$ and (1.3) holds, i.e. the stopping problem is in the weakly monotone case.

Proof: It is enough to show that for all $t \in T$ $B_t = C_t$ holds.

We have

$$f(Z_t,t)1_{C_t} \geq E(f(Z_{t+h},t+h)\,|\,A_t)1_{C_t} = E(f(Z_{t+h},t+h)\,|\,Z_t)1_{C_t},$$

thus

$$Af(Z_t,t) \cdot 1_{C_t} \leq 0,$$

which implies $C_t \subset B_t$.

By using (1.1) for any $h \in T$

$$E(f(Z_{t+h},t+h)\,|\,A_t) = f(Z_t,t) + E(\int_t^{t+h} Af(Z_s,s)ds\,|\,A_t)$$

thus

$$E(f(Z_{t+h},t+h)\,|\,A_t) \cdot 1_{B_t} = f(Z_t,t) \cdot 1_{B_t} + E(1_{B_t} \cdot \int_t^{t+h} Af(Z_s,s)ds\,|\,A_t) \leq f(Z_t,t) \cdot 1_{B_t},$$

which yields $B_t \subset C_t$.

In the following we will formulate the monotone case for general stochastic processes and show that (1.4) can also be obtained then.

2. The general case

We now turn to the case of a general $(X_t)_{t \in T}$. To define the monotone case we consider for all $t, h \in T$ fixed versions X_t^h of $E(X_{t+h}\,|\,A_t)$ (note $X_t^o = X_t$) and assume that for every $t \in T$ the stochastic process $(X_t^h)_{h \in T}$ has a.s. right-continuous paths and that the limit

$$\lim_{h \downarrow 0} (X_t^h(\omega) - X_t^o(\omega))/h$$

exists for a.a. ω.

We denote this limit by $H_t(\omega)$ and thus obtain a real-valued stochastic process $(H_t)_{t \in T}$.

Analogously to (1.2) we define the stopping problem to be in the monotone case if the following holds:

(2.1) $\{H_t<0\}\subset\{H_{t+h}<0\}$ for all $t,h\in T$ and $\bigcup_{t\in T}\{H_t<0\}=\Omega$.

Then we can prove the following result:

(2.2) Theorem: Assume (2.1) and that for all $t\in T$

(i) $(X_t^h)_{h\in T}$ has a.s. paths which are absolutely continuous.

(ii) $\{(X_t^h-X_t^o)^+/h:\ 0<h<c\}$ is uniformly integrable for some $c>0$.

Then $\{H_t<0\}=C_t$ for all $t\in T$ and the stopping problem is in the weakly monotone case.

Proof: Fix $t\in T$. Obviously $C_t\subset\{H_t<0\}$ so it only remains to show the reverse inclusion. This will be done by contradiction, so let us assume

$$P(\{H_t<0\}\cap\{X_t<\operatorname*{ess\,sup}_{h>0}E(X_{t+h}\mid A_t)\})>0.$$

Thus there exists some $h>0$ such that

$$P(\{H_t<0\}\cap\{X_t^o<X_t^h\})>0.$$

We define a measurable stochastic process $(DX_t^s)_{s\in T}$ by

$$DX_t^s(\omega)=\liminf_{h\downarrow 0}\,(X_t^{s+h}(\omega)-X_t^s(\omega))/h$$

and by assumption (i) we have for a.a. ω

$$X_t^h(\omega)-X_t^o(\omega)=\int_0^h DX_t^s(\omega)ds.$$

Defining $A=\{H_t<0\}\cap\{X_t^o<X_t^h\}$ we obtain

$$\lambda(\{s<h:\ DX_t^s(\omega)>0\})>0 \quad\text{for a.a. } \omega\in A$$

and thus

$$\int_A\lambda(\{s<h:\ DX_t^s(\omega)>0\})dP>0.$$

By Fubini's theorem

$$\int_0^h P(\{\omega\in A:\ DX_t^s(\omega)>0\})ds>0,$$

which implies the existence of some $s>0$ and some $\varepsilon>0$ such that

$$P(\{H_t<0\}\cap\{DX_t^s>2\varepsilon\})>0.$$

Thus there exists some $\delta>0$ such that we have

$$P(\{H_t < 0\} \cap \{\inf_{0<h<\delta} (X_t^{s+h} - X_t^s)/h > \varepsilon\}) > 0.$$

We denote the above set by G and it follows

$$1_G\,(E(X_{t+s+h} \mid A_t) - E(X_{t+s} \mid A_t)) > \varepsilon h\, 1_G \text{ for all } 0<h<\delta$$

thus (note $G \in A_t$)

$$E(1_G(E(X_{t+s+h} \mid A_{t+s}) - X_{t+s}) \mid A_t) > \varepsilon h\, 1_G$$

and

$$E(1_G(E(X_{t+s+h} \mid A_{t+s}) - X_{t+s})) > \varepsilon h\, P(G),$$

i.e.

$$E(1_G(X_{t+s}^h - X_{t+s}^o)/h) > \varepsilon\, P(G) \text{ for all } 0<h<\delta.$$

By assumption (ii) we may apply Fatou's lemma and obtain

$$E(1_G \lim_{h \downarrow 0} (X_{t+s}^h - X_{t+s}^o)/h) > \limsup_{h \downarrow 0} E(1_G(X_{t+s}^h - X_{t+s}^o)/h) > \varepsilon P(G) > 0,$$

thus

$$E(1_G\, H_{t+s}) > 0.$$

But this contradicts $G \subset \{H_t < 0\} \subset \{H_{t+s} < 0\}$.

The infinitesimal look-ahead rule here takes the form

$$\rho = \inf\{t \in T: H_t < 0\},$$

and it follows from the theorem, that under suitable conditions ρ and the stopping time σ of the weakly monotone case also coincide in the general situation. We can thus use the results of Irle [3] on optimality of σ to obtain optimality of ρ in the monotone case.

As can be seen already from simple deterministic examples, the condition on absolute continuity may not be omitted in general: Consider first $X_t = 1_{[1,\infty)}(t)$. Then (2.1) holds with $\{H_t < 0\} = \Omega$ for all t, but $\rho = 0$ yields smallest reward. Here $C_t = \emptyset$ for $t<1$ and $C_t = \Omega$ for $t>1$ so that (1.3) holds, and $\sigma = 1$ is an optimal stopping time. For $X_t = \sum_{n=0}^{\infty} n\, 1_{[n,n+1)}(t)$ again $\{H_t < 0\} = \Omega$ for all t and $\rho = 0$, but now $C_t = \emptyset$ for all t.

3. Application

We consider a mathematical model for the following situation:

A certain plant yields a random profit $Z_{t+h} - Z_t$ when in operation from time t to time t+h. This plant breaks down at random time points, which follow a Poisson process, and then has to be repaired with random costs of repair Y_i when the i - th break-down has occured.

Thus assume that on the basic complete probability space are given:

A stochastic process $(Z_t)_{t \in T}$ with independent increments having right-continuous and increasing paths.

A Poisson process $(N_t)_{t \in T}$ and a sequence $(Y_n)_{n \in \mathbb{N}}$ of independent random variables, $Y_n > 0$ for all n, with costs of repair

$$R_t = \sum_{i=1}^{N_t} Y_i .$$

We may then ask the question how long it is worth-while to operate this plant, which leads to the problem of optimal stopping for

$$X_t = Z_t - R_t$$

with respect to the usual right-continuous filtration induced by the stochastic process $(Z_t, R_t, N_t)_{t \in T}$.

The following assumptions will be made:

$(Z_t)_{t \in T}$, $(N_t)_{t \in T}$ and $(Y_n)_{n \in \mathbb{N}}$ are stochastically independent.

$c(t) = EZ_t$ is differentiable and concave,

$f(i) = EY_i$ is increasing, such that

$$\sup_i f(i) > \inf_t c'(t).$$

We can now compute:

$$E(X_{t+h} \mid A_t) = Z_t + E(Z_{t+h} - Z_t) - R_t - E\Big(\sum_{i=N_t+1}^{N_{t+h}} Y_i \mid A_t \Big)$$

$$= X_t + c(t+h) - c(t) - E\Big(\sum_{i=N_t+1}^{N_{t+h}} f(i) \mid N_t \Big)$$

$$= X_t + c(t+h) - c(t) - \sum_{k=1}^{\infty} \sum_{i=1}^{k} f(N_t+i) e^{-h} h^k/k!$$

Assuming convergence of the infinite series we obtain

$$\frac{d}{dh} X_t^h = c'(t+h) - \sum_{k=1}^{\infty} \sum_{i=1}^{k} f(N_t+i)e^{-h}((-1)h^k/k!+h^{k-1}/(k-1)!)$$

and

$$H_t = c'(t) - f(N_t+1),$$

thus

$$\{H_t < 0\} = \{f(N_t+1) > c'(t)\}.$$

From the assumptions

$$\{H_t < 0\} \subset \{H_{t+h} < 0\} \quad \text{and} \quad \bigcup_{t \in T} \{H_t < 0\} = \Omega$$

so that by (2.2) we are in the weakly monotone case with

$$\rho = \sigma = \inf\{t \in T: N_t > f^{-1}(c'(t))-1\},$$

where f^{-1} denotes the generalized inverse of f. For large t $f^{-1}(c'(t))$ is bounded so that $E\rho^n < \infty$ holds for all $n \in \mathbb{N}$.

We will next show that ρ is optimal in the following sense:

(3.1) $EX_\rho \geq EX_\tau$ for all stopping times τ with
$EZ_\tau < \infty$ or $ER_\tau < \infty$,

which includes all τ with $E\tau < \infty$.

This follows from Irle [3], if we can prove validity of the next condition (3.2):

(3.2) For any increasing sequence $(\tau_n)_n$ of stopping times
with $\tau_n < \sigma$ we have $EX_{\sup \tau_n} \geq \limsup EX_{\tau_n}$.

So consider such a sequence $(\tau_n)_n$, $\tau = \sup \tau_n$. It follows from monotonicity

$$\lim EZ_{\tau_n} \leq EZ_\tau \leq EZ_\rho = Ec(\rho) < \infty,$$

since $c(t) \leq \gamma_1 t + \gamma_2$ for some constants γ_1, γ_2.

Furthermore the Poisson process $(N_t)_{t \in T}$ is quasi-left-continuous with respect to the filtration $(A_t)_{t \in T}$ (see e.g. Blumenthal and Getoor [2], I.12), thus $N_\tau = \lim N_{\tau_n}$, which implies $N_\tau = N_{\tau_n}$ for n large enough and thus $R_\tau = R_{\tau_n}$ for such n. This yields $ER_\tau = \lim ER_{\tau_n}$, so that

$$EX_\tau = EZ_\tau - ER_\tau \geq \lim EZ_{\tau_n} - \lim ER_{\tau_n} = \lim EX_{\tau_n} .$$

For an extension of the optimality of ρ in (3.1) assume that the following condition holds

(3.3) $$E \sup_{t \in T} (\alpha Z_t - R_t) < \infty \quad \text{for some} \quad \alpha > 1.$$

Then under (3.3) for any stopping time

$$EX_\tau^+ < \infty \text{ and } EX_\tau \leq E \sup_{t \in T} (\alpha Z_t - R_t) - (\alpha - 1) EZ_\tau ,$$

thus $EX_\tau = -\infty$ if $EZ_\tau = \infty$, which implies the optimality of ρ among all stopping times τ.

Let us remark that this application could also be treated within the Markov framework by choosing an appropriate space-time Markov process but that the general framework yields a natural approach for this problem.

References.

[1] Athreya, K.B. and Kurtz, T.G. (1973). A generalization of Dynkin's identity and some applications. Ann. Prob. 1, 570-579.

[2] Blumenthal, R. and Getoor, R. (1968). Markov Processes and Potential Theory. Academic Press, New York.

[3] Irle, A. (1979). Monotone stopping problems and continuous time processes. Z. Wahrscheinlichkeitstheorie verw. Gebiete 48, 49-56.

[4] Rosenkrantz, W.A. (1974). An application of the Hille-Yosida-theorem to the construction of martingales. Indiana Univ. Math. J. 24, 527-532.

[5] Ross, S.M. (1971). Infinitesimal look-ahead stopping rules. Ann. Math. Statist. 42, 297-3o3.

[6] Shapiro, C.P. and Wardrop, R.L. (198o). Dynkin's identity applied to Bayes sequential estimation of a Poisson process rate. Ann. Statist. 8, 171-182.

NUMERICAL INVESTIGATION OF THE TWO ARMED BANDIT

by

P. W. Jones and H. A. Kandeel

Department of Mathematics, University of Keele,
Keele, Staffordshire, ST5 5BG, U.K.

1. Introduction

This paper is concerned with Bernoulli two armed bandits with independent beta priors for the unknown success probabilities where there are a finite number of trials, N, and the objective is to maximise the overall expected return. The two armed bandit with one probability known is also considered.

The results of numerical investigations into the structure of the optimal design, obtained by dynamic programming, are reported for the cases where the rewards are undiscounted, they suggest a modification of the one step ahead design [Jones (1975)] which gives a greater efficiency than the original. Implications for the display of the optimal design are discussed. The effect of discounting and the measurement of uncertainty are considered and some two stage designs are presented.

2. Two Armed Bandit

2.1 Undiscounted rewards

Consider first the two armed bandit where there is no discounting. Suppose that the unknown probability of success, p_i, on arm i is assigned a beta prior with parameters a_i, b_i (beta (a_i, b_i)) $1 \leq a_i < b_i$, they are assumed for convenience to be integers, the implication of this assumption is discussed later, then after c_i successes on d_i trials on arm i, the posterior distribution of p_i is beta (r_i, n_i) with $r_i = a_i + c_i$, $n_i = b_i + d_i$ with density proportional to

$$p_i^{r_i-1} (1-p_i)^{n_i-r_i-1}, \quad i = 1, 2 \qquad r_i < n_i .$$

It is also assumed that the p_i's are independent.

The optimal decision at each point (r_1, n_1, r_2, n_2) in four dimensional integer space is found by using the well known dynamic programming equations.

Suppose that $S(r_1, n_1, r_2, n_2)$ is the maximum expected return of the remaining $N-d_1-d_2$ trials and that $S_i(r_1, n_1, r_2, n_2)$ is the maximum expected return given that the next trial takes place on arm i, i=1, 2 then

$$S = \max [S_i]$$

and

$$S_1(r_1, n_1, r_2, n_2) = r_1/n_1 \; [1+S(r_1+1, n_1+1, r_2, n_2)] + \left(\frac{n_1-r_1}{n_1}\right) S(r_1, n_1+1, r_2, n_2),$$

with S_2 similarly defined. The design is then found by working back from $(r_1, n_1$ $r_2, n_2)$, with $(n_1+n_2) = N+b_1+b_2$ where $S(r_1, n_1, r_2, n_2) = 0$.

Due to the backward progress of the computation it is not known which points are reachable by any simple path starting at (a_1, b_1, a_2, b_2) until this origin is reached. It is also necessary to recompute the design for different N and different priors, although it is possible to recover the design for smaller N for other integer parameter priors, this is obviously not generally true where the priors do not have integer parameters. It is therefore worthwhile to look at the structure of the design to help reduce the computation and to solve the related problem of how it should be displayed.

The following four results help achieve a reduction in complexity.

Theorem 1 $S(r_1, n_1, r_2, n_2) = S(r_2, n_2, r_1, n_1)$

Proof by induction on (n_1+n_2).

The theorem implies that different decisions are made at the two symmetric points.

Corollary 1 (r, n, r, n) is an indifference point

Proof follows directly from Theorem 1.

Remark The above two results depend on the assumption of integer priors, or priors with the same parameters, and will obviously not be generally true for non-integer priors.

Theorem 2 If (r_1, n_1, r_2, n_2) is a continue (pull) 1 point then (r_1+1, n_1, r_2, n_2) is a continue 1 point with a similar result for continue 2 points.

This was conjectured in Jones (1976) and proved by Kalin and Theodorescu (1981).

Theorem 3 [Theorem 6.2, Berry (1972)] If (r_1, n_1, r_2, n_2) is a continue 1 point then (r_1+1, n_1+1, r_2, n_2) is a continue 1 point, with a similar result for continue 2 points (the stay on a winner rule).

To infer what other properties the optimal design might possess, it is instructive to consider the suboptimal one step ahead (OSA) design, that is: at point (r_1, n_1, r_2, n_2)

if $r_1/n_1 > r_2/n_2$ continue 1

if $r_1/n_1 < r_2/n_2$ continue 2

otherwise the point is an indifference point.

Note that the OSA design possesses all the above properties of the optimal. Numerical results, some of which are presented in Table 1 suggest that this design is nearly optimal which implies that there are a large number of points where the two designs give the same decision. That is where the future expected returns in the dynamic programming equations are not important, they introduce uncertainty only at a few points.

TABLE I

Optimal expected returns and efficiencies of OSA and modified OSA for N = 5(5)50 and two sets of priors. [Some results from Jones (1975)]

N	optimal expected return	(1,2)v(1,2) efficiency of OSA (%)	efficiency of modified OSA (%)	optimal expected return	(1,2)v(1,3) efficiency of OSA (%)	efficiency of modified OSA (%)
5	2.8889	99.71	100	2.5917	100	100
10	6.0218	99.48	99.97	5.3261	99.84	99.98
15	9.2115	99.32	99.84	8.1069	99.71	99.91
20	12.4313	99.19	99.71	10.9145	99.59	99.80
25	15.6696	99.08	99.59	13.7391	99.48	99.69
30	18.9199	98.99	99.48	16.5746	99.39	99.50
35	22.1799	98.90	99.39	19.4190	99.30	99.52
40	25.4469	98.83	99.31	22.2700	99.23	99.44
45	28.7195	98.77	99.23	25.1264	99.16	99.37
50	31.9967	98.71	99.16	27.9872	99.10	99.31

It is obvious that the main difficulties occur where the OSA design gives an indifference point $r_1/n_1 = r_2/n_2$, $n_1 \neq n_2$, this is where the comparison between the future expected returns only are used to determine the decision in the optimal design. From numerical work it is clear that at these OSA indifference points the optimal design produces a continue 1 point if $n_1 < n_2$ and a continue 2 point if $n_1 > n_2$ except at $(n_1+n_1) = (N-1) + b_1 + b_2$ where the optimal and OSA designs coincide. If the OSA design is now modified to take account of this tie breaking property then a significant increase in efficiency follows, this is also illustrated in Table I. There are now very few points at which the modified OSA and optimal designs give differing decisions. It seems reasonable to conjecture that the optimal design possesses this property.

To display the optimal design it is now only necessary to give those points which do not give coincident decisions with the modified OSA (or any other simple suboptimal design). The closeness to the modified OSA also helps solve the problem of whether the points are reachable by any sample path.

Example 1 Consider N = 10 and beta (1, 2) (uniform) priors for p_i then the reachable points where the optimal design gives a different decision to modified OSA are (4, 7, 1, 2), (3, 8, 1, 3), (5, 9, 1, 2) all continue 2 points (and their symmetric points).

It may be noted from Table I that the efficiency of the OSA design decreases as N becomes larger, further, it was observed that the expected returns of the optimal and OSA designs were approximately linear in N for $25 \leq N \leq 50$ for a wide range of priors, some values of the constant A and slope B are presented in Table II. The difference between the slopes for the two designs gives some indication of the asymptotic loss per observation from using the subpotimal design, this suggests that the OSA is not asymptotically optimal. A modification of the OSA that ensures its asymptotic optimality is given in Bather (1981).

TABLE II

Values of slope (B) and constant (A) in approximate linear relationship $S(0,0,0,0) = A + BN$ for different priors. (Values for OSA in brackets).

Priors	A	B
(1, 2)v(1, 2)	-0.6753 (-0.5427)	0.6532 (0.6424)
(1, 2)v(1, 3)	-0.5255 (-0.4092)	0.5700 (0.5627)
(1, 2)v(1, 4)	-0.3788 (-.3823)	0.5386 (.5362)
(1, 2)v(1, 5)	-0.2731 (-.1342)	0.5237 (.5190)

For large N and for ease of use, two stage designs may be constructed where the first stage consists of a sequential design and the second consists of using the 'best' arm only. This kind of design could be used where the optimisation problem is combined with a terminal decision. This approach may be suitable in the analysis of clinical trials, however it has been noted in several papers that a terminal decision in this context is perhaps inappropriate.

A brief investigation was carried out for two types of design:

Type A: fixed first stage sample size N_1 with the optimal used up to that point and the arm with the larger r_i/n_i used for the remaining $N-N_1$ trials.

Type B: variable first stage sample size, where the optimal design is used provided $|r_1/n_1 - r_2/n_2| < \delta$, where δ is preassigned, otherwise continue 1 or continue 2 if $r_1/n_1 >$ or $< r_2/n_2$ respectively. Some numerical results are presented in Table III.

Type A designs produce a fairly efficient design even in cases where N_1 is small compared to N. It is surprising that there is so little loss in efficiency in these cases, however, this design could appear to be rather inefficient if it was judged under other criteria such as minimising the number of trials using the poorer arm.

The expected returns of type B designs obviously approach those of the optimal as $\delta \to 1$ since for large δ the points satisfying the condition are not reachable, the value of δ must be obtained by balancing the loss in efficiency against the ease of use. To further the ease of use in these types of design, the optimal design in the first stage could be replaced by the OSA or play the winner rule.

TABLE III

Efficiencies of two stage designs

Type A first stage N_1 = 10 (1, 2) v (1, 2) priors

$N-N_1$ =	10	20	30	40
Efficiency (%) =	99.71	99.30	98.97	98.73

Type B (1, 2) v (1, 2) priors

N =	4	5	6	7	8	9	10
δ = .1	98.78	98.08	97.39	96.81	96.30	95.89	95.49
.2	100	100	99.92	99.83	99.73	99.64	99.51
.3	100	100	100	99.97	99.93	99.89	99.84
.4	100	100	100	100	100	100	99.997

2.2 Discounted rewards

If rewards one trial away are discounted at a constant rate, α, then the optimal design is obtained by modifying the dynamic programming equations above by defining

$$S(r_1, n_1, r_2, n_2) = r_1/n_1[1+\alpha\ S(r_1+1, n_1+1, r_2, n_2)] + \left(\frac{n_1-r_1}{n_1}\right)\alpha S(r_1, n_1+1, r_2, n_2)$$

similarly for S_2.

Intuitively it would be expected that the OSA design would be closer to the optimal as $\alpha \to 0$ since the uncertainty effect represented by the future expected returns would diminish, also due to the dominance of the (r_i/n_i) terms there would be fewer points like those in Example 1 which are not predictable using any sub-optimal design.

Example 2 For N = 10 and (1, 2) priors for the p_i, the points at which the optimal design differed from the modified OSA were obtained for different discount factors with the following results:

α =	.9	.75	.5
	(4, 7, 1, 2)	(5, 9, 1, 2)	none
	(5, 9, 1, 2)		
	(3, 8, 1, 2)		

and symmetric points.

The finite horizon Gittins index [Gittins and Jones (1979)] was calculated for each of the three cases, it gave the optimal design for α = .9 but also gave the design with the same three points as α = .9 for α = .75, .5.

3. Two armed bandit and probability known

If one probability is known, say $p_2 = p$ and p_1 is assigned to beta (a, b) prior then the problem becomes one of optimal stopping for arm 1 and the dynamic programming equations for the discounted case becomes

$$S(r, n) = \max [S_1(r, n), (N-n+b)p]$$

with

$$S_1(r, n) = r/n\ [1+S(r+1, n+1)] + \left(\frac{n-r}{n}\right) S(r, n+1)$$

where S(r, n) is the maximum expected return at point (r, n) in 2 dimensional space and $S_1(r, n)$ is the maximum expected return when arm 1 is pulled next, for sample paths starting at (a, b). When the rewards are discounted at a constant rate α, the equations become

$$T(r, n) = \max \left[T_1(r, n), \left(\frac{1-\alpha^{N-n+b}}{1-\alpha}\right) p\right]$$

with

$$T_1(r, n) = r/n[1+\alpha\ T(r+1, n+1)] + \left(\frac{n-r}{n}\right) \alpha T\ (r, n+1).$$

The effect of introducing a discount factor is to produce a design which stops for smaller values of n for fixed r than the undiscounted design, this may easily be shown to be true by proving the following theorem.

Theorem 4

$$T_1(r, n) \Big/ \left[\frac{1-\alpha^{N-n+b}}{1-\alpha}\right] \leq S_1(r, n) \Big/ (N-n+b)$$

Proof by induction, noting that T(r, n), S(r, n) are increasing functions of n for fixed r and confining attention to (r, n) such that $r/n < p < (r+1)/(n+1)$.

To measure the reduction in uncertainty introduced by the discount factor, Gittins and Jones (1979) suggested the value of Δ where $(r+\Delta)/n \leq p$ on the stopping boundary, again this gives the departure from the OSA design. It was shown in Jones (1978) in the undiscounted case that if (r, n) is a stopping point $r/n < p$ (Hengartner, Kalin and Theodorescu have pointed out that this not iff) for $n < N+b-1$, this proof may easily be extended to the discounted case, from which it follows that $\Delta > 0$. Some maximum and minimum values of Δ are given in Table IV for N = 100.

The values of Δ tend to decrease with increasing N keeping α, p constant, the maximum value of Δ increases with increasing α or p keeping the other two constant. The smaller values of Δ, as expected, occur near $n = N+b-1$. A consequence of decreasing α is that it produces designs where arm 1 is never pulled.

TABLE IV

Maximum and minimum values of Δ for N = 100 and different discount factors and known probability p and (1, 2) prior for p_1.

p =	.1	.5	.7
α = .75	.1, 1.2	.5, 1	pull 2 only
.9	.1, 1.4	.5, 1.5	.3, 1.7
undiscounted	.1, 2	.5, 2.5	.3, 2.8

4. Discussion

The structure of the optimal design for maximising the sum of observations in the two armed bandit has been discussed. It would perhaps be useful to measure the efficiency of the design produced under other criteria such as the number of trials on the poorer arm (a treatment in a clinical trial), or the probability of correctly selecting the better arm. For some applications it might be better to treat the problem as a two decision problem under loss functions depending on (p_1-p_2). The results of simulation studies comparing designs produced under different criteria will be presented elsewhere.

Acknowledgment

The paper has been improved by the helpful and stimulating comments of participants at the conference.

References

Bather, J. A. (1981). Randomised allocation of treatments in sequential experiments, (with discussion). Journ. Roy. Statist. Soc. B, 43, 265-292.
Berry, D. A. (1972). A Bernoulli two armed bandit. Ann. Math. Statist., 43. 871-897.
Gittins, J. C. (1979). Bandit processes and dynamic allocation indices, (with discussion). Journ. Roy. Statist. Soc. B, 41, 148-177.
Gittins, J. C. and Jones, D. M. (1979). A dynamic allocation index for the discounted multiarmed bandit problem. Biometrika, 66, 561-5.
Hengartner, W., Kalin, D. and Theodorescu, R. (1981). On the Bernoulli two armed bandit problem. Math. Operationsforsch. Statist., 12, 307-316.
Jones, P. W. (1975). The two armed bandit. Biometrika, 62, 523-4.
Jones, P. W. (1976). Some results for the two armed bandit problem. Math. Operationsforsch. Statist., 7, 471-475.
Jones, P. W. (1978). On the two armed bandit with one probability known. Metrika, 25, 235-9.
Kalin, D and Theodorescu, R. (1981). Abstract 81t-80. Bull. Inst. Math. Statist., 10, 5, 224. (See also unpublished research report from the authors).

UNIFORM BOUNDS FOR A DYNAMIC PROGRAMMING MODEL UNDER ADAPTIVE CONTROL USING EXPONENTIALLY BOUNDED ERROR PROBABILITIES

by Michael Kolonko,
Karlsruhe

1. Introduction

We consider a control model of dynamic programming. The distribution of the disturbance variable depends on an unknown parameter $\theta \in \Theta$ which is estimated by T_n. We give a bound on how much we loose compared with the maximum total reward if we use a plan that would be optimal if the estimated value T_n was the true value of θ. We generalize a result of Rényi for finite Θ to a discretized infinite Θ to give a bound that holds independently of the true value of θ and that tends to zero exponentially with growing sample size n.

2. The Model

We consider the following model of dynamic programming :

- S is the measurable state space;
- A is the measurable action space;
- W is the measurable disturbance space;
- $D \subset S \times W \times A$ is a measurable set containing the graph of a measurable function $S \times W \to A$. $D(s,w) := \{a| \ (s,w,a) \in D\}$ is the set of admissible actions.
- $p^o \in \mathbb{P}(S)$ is the starting distribution, here $\mathbb{P}(S)$ denotes the set of all probability measures on S.
- $p : D \to \mathbb{P}(S)$ is a transition probability, i.e. $p(\cdot|s,w,a) \in \mathbb{P}(S)$ and $p(B|\cdot,\cdot,\cdot)$ is measurable for all measurable $B \subset S$.
- Θ is the metric parameter space endowed with the Borel σ-algebra. $|\cdot,\cdot|$ denotes the metric on Θ.
- $\mu : \Theta \to \mathbb{P}(W)$ is a transition probability. Let $l(\cdot,\theta)$ denote the density of μ_θ with respect to some σ-finite measure σ on W.
- $\nu \in \mathbb{P}(\Theta)$ is the apriori distribution ;
- $r : D \times \Theta \to \mathbb{R}$ is the measurable one-step reward function, r is absolutely bounded by $\bar{r} \in \mathbb{R}$.

- $\beta \in (0,1]$ is the <u>discount factor.</u>

This model describes the following control situation:
at fixed points of time 1,2,.. a system is observed to be in one of the states $s \in S$. A disturbance occurs according to the distribution μ_θ, where θ is unknown with an apriori distribution ν. Then, using a decision rule f, the controller chooses an action $a = f(s,w) \in D(s,w)$. The next state s' is distributed according to $p(\cdot|s,w,a)$. From this transition a reward $r_\theta(s,w,a)$ is earned which is discounted back to time 1 by the discount factor β.The aim of the controller is to maximize the total expected reward over a fixed number of N periods.
We assume that the controller knows how to choose the actions optimally if θ is known to him. As θ is unknown, he estimates it by T_n and then behaves as if T_n was the true value. The amount of total reward he looses compared with the maximum is called regret.
It is the aim of this paper to give a bound for this regret which holds uniformly for all apriori distributions satisfying a mild condition (see below).

As usual, let $F := \{f : S \times W \to A \mid f \text{ measurable, } f(s,w) \in D(s,w)\}$ denote the set of all decision rules. F^∞ is the set of all Markov plans δ. Plans $\delta = (f,f,..)$ with $f \in F$ are called stationary.
For each $\delta = (f_1,f_2,..) \in F^\infty$ there is a probability measure $P = P_{\nu,\delta}$ on $\Omega := \Theta \times (S \times W)^\infty$ such that

$$P(T \in G) = \nu(G),$$
$$P(X_1 \in B,\ Y_1 \in C \mid T = \theta) = p^o(B)\ \mu_\theta(C),$$
$$P(X_{m+1} \in B,\ Y_{m+1} \in C \mid X_1,Y_1,..X_m,Y_m,T = \theta) = p(B|X_m,Y_m,A_m)\ \mu_\theta(C) \text{ and}$$
$$P(T \in G \mid X_1,Y_1,..X_m,Y_m) = P(T \in G \mid Y_1,..Y_m)$$

for B,C,G chosen appropriately. Here T,X_m,Y_m are the projections on Θ, the m-th factor S and the m-th factor W of Ω. $A_m := f_m(X_m)$. Hence T,X_m, Y_m and A_m are the parameter, the m-th state, disturbance and action variable.
Note that (Y_m) has a marginal distribution P_ν independent of δ. Given $T = \theta$, (Y_m) is a conditionally independent sequence with distribution μ_θ. Moreover, given $(Y_1,Y_2,..Y_m)$, $(X_1,X_2,..X_m)$ is conditionally independent of T, hence $(Y_1,Y_2,..Y_m)$ may be regarded as a sufficient statistic for estimating θ on stage m.
Putting $h = (s,w)$, we define the total expected reward over horizon $N \in \overline{\mathbb{N}}$ for starting in h under plan $\delta \in F^\infty$ and given $T = \theta$ as

$$V_\theta^N(h,\delta) := E_{\nu,\delta}\left[\sum_{k=1}^N \beta^{k-1}\, r_\theta(X_k,Y_k,A_k) \mid (X_1,Y_1) = h,\ T = \theta\right].$$

If we assume $\beta < 1$ for $N = \infty$, V^N is well defined.

We make the following general assumptions about optimal plans for known θ (i.e. given $T = \theta$) :

For each $\theta \in \Theta$ and $n \in \overline{\mathbb{N}}$ there is $f_n(\cdot,\theta) \in F$ such that

(i) $(s,\theta) \to f_n(s,\theta)$ is measurable,

(ii) for all $N < \infty$ $\delta^N(\theta) := (f_N(\cdot,\theta), f_{N-1}(\cdot,\theta),\ldots f_1(\cdot,\theta), f_{N+1},\ldots)$ is an optimal Markov plan (given $T = \theta$) with $f_n \in F$ arbitrary, $n<N$,

(iii) for $N = \infty$ $\delta^\infty(\theta) := (f_\infty(\cdot,\theta), f_\infty(\cdot,\theta),\ldots)$ is an optimal stationary plan (given $T = \theta$).

From (i) it follows that $(\theta,\theta',s,w) \to V_\theta^N(s,w,\delta^N(\theta'))$ is measurable.
(ii) and (iii) imply that for all $\theta \in \Theta$, $N \in \overline{\mathbb{N}}$ and $h \in S \times W$

$$\sup_{\delta \in F^\infty} V_\theta^N(h,\delta) = V_\theta^N(h,\delta^N(\theta)) =: V_\theta^N(h) .$$

3. Lipschitz Conditions

In order to be able to give a bound on the effect the estimation error connected with T_n has on the total rewards, we have to impose the following Lipschitz conditions :

(C) There are constants $\rho>0$, $\eta>0$ such that for all $(s,w,a) \in D$ and all $\theta,\theta' \in \Theta$

$$|r_\theta(s,w,a) - r_{\theta'}(s,w,a)| \le \rho|\theta,\theta'| \quad \text{and}$$
$$\|\mu_\theta - \mu_{\theta'}\| \le \eta|\theta,\theta'|$$

where $\|\cdot\|$ denotes the total variation of the signed measure $\mu_\theta - \mu_{\theta'}$.

The following Theorem shows that this Lipschitz continuity carries over to the total rewards if we put

$$\rho_N := \sum_{k=1}^N \beta^{k-1} \left(\rho + \beta\eta\frac{1-\beta^{N-k}}{1-\beta}\right) \qquad \text{for } N \in \overline{\mathbb{N}}$$

THEOREM 1

Let (C) hold, then for $N \in \overline{\mathbb{N}}$, $\theta,\theta' \in \Theta$

$$\sup_h |V_\theta^N(h,\delta^N(\theta')) - V_\theta^N(h)| \le 2\,\rho_N|\theta,\theta'| ;$$

i.e. if θ is the true parameter, but plan $\delta^N(\theta')$ (optimal for parameter θ') is used, then the regret is bounded by $2\rho_N|\theta,\theta'|$.

For a proof of this Theorem cp. Kolonko(82), Theorem 3.2.

4. A Bayes Estimator with Exponentially Bounded Error Probability

Now we are going to define an estimator T_n such that Theorem 1 with θ' replaced by T_n yields a reasonable result. Although Θ is allowed to be an infinite metric space we have to assume some finite structure in the dependence of $\theta \to \mu_\theta$. We assume that for some given $\varepsilon > 0$ the following condition holds :

(R_ε) There is $K \in \mathbb{N}$ and a partition $\Theta = \sum_{i=1}^{K} \Theta_i$ of Θ into sets Θ_i such that

(1) $\sup \{|\theta,\theta'| \mid \theta,\theta' \in \Theta_i\} \leq \varepsilon$, for all $1 \leq i \leq K$;

(2) (Y_n) are conditionally independent given $T \in \Theta_i$, $1 \leq i \leq K$ and

(3) the conditional distribution of Y_n given $T \in \Theta_i$ is different for different i, $1 \leq i \leq K$.

Note that (2) strengthens the property of conditional independence of (Y_m) given $T = \theta$.

Put

(4) $Q_j(w_1,w_2,..w_n) := P_\nu(T \in \Theta_j | Y_m = w_m, 1 \leq m \leq n)$, $w_m \in W, 1 \leq m \leq n$ and

(5) $l(w,\Theta_j) := \nu(\Theta_j)^{-1} \int_{\Theta_j} l(w,\theta)\nu(d\theta)$

$l(w,\Theta_j)$ is the conditional density of Y_m given $T \in \Theta_j$.

We want to estimate the true Θ_j, i.e. that index j with $T \in \Theta_j$. We use the following Bayes estimator $T_n = T_n(w_1,w_2,..w_n)$

(6) $T_n(w_1,..w_n) = j :\Leftrightarrow Q_j(w_1,..w_n) = \max^*_{1 \leq i \leq K} Q_i(w_1,..w_n)$

where $\max^*$ means that j is chosen to be the largest maximizer in case there are more than one.

This set-up is a straight generalisation of a finite parameter space estimation problem considered in Rényi (64), cp. also Vincze (67). There it was shown that the error probability converges exponentially to zero. This result also holds in our case :

THEOREM 2

Let (R_ε) hold for some given $\varepsilon \geq 0$ and assume

(7) $\nu(\Theta_i) > 0$ for $1 \leq i \leq K$.

Then

(8) $q := \max \{\inf_{0<\alpha<1} \int_W l(w,\Theta_j)^\alpha \, l(w,\Theta_i)^{1-\alpha} \sigma(dw) \mid 1 \leq i,j \leq K, i \neq j\} < 1$

and

$$P_\nu(T \notin \Theta_{T_n}) \leq (K-1)\, q^{n-1} .$$

Note that the bound holds uniformly for all apriori distributions satisfying (7), including e.g. $\nu(\Theta_i) = K^{-1}$ $(1 \le i \le K)$.

The proof follows the lines of Rényi (64) and is only sketched here. First we note that $l(\cdot,\Theta_j), l(\cdot,\Theta_i)$ are different on a set of positive σ-measure for different i,j by (3). Hence one may apply the strict Hölder inequality to obtain

$$q_{ij}^{\alpha} := \int_W l(w,\Theta_j)^{\alpha}\, l(w,\Theta_i)^{1-\alpha}\, \sigma(dw) < 1$$

for all $0<\alpha<1$ and $i \neq j$. Hence $q < 1$.

$\prod_{m=1}^{n} l(w_m,\theta)$ is the common density of $Y_1, Y_2, \ldots Y_n$ with respect to σ^n given $T = \theta$. From (2) it follows by straightforward calculations that

$$\int_{\Theta_i} \prod_{m=1}^{n} l(w_m,\theta)\, \nu(d\theta) = \nu(\Theta_i)^{1-n} \prod_{m=1}^{n} \int_{\Theta_i} l(w_m,\theta)\, \nu(d\theta) .$$

Now using (6) and the Markov inequality we see that for $\theta \in \Theta_i$

$$P_\nu(T_n = j \mid T = \theta) \le P_\nu(Q_j(Y_1,\ldots Y_n) \ge Q_i(Y_1,\ldots Y_n) \mid T = \theta)$$
$$\le E_\nu\left[\left(\frac{Q_j(Y_1,\ldots Y_n)}{Q_i(Y_1,\ldots Y_n)}\right)^{\alpha} \mid T = \theta\right]$$

Hence

$$P_\nu(T_n = j\, , \, T \in \Theta_i) \le \int_{\Theta_i} E_\nu\left[\left(\frac{Q_j(Y_1,\ldots Y_n)}{Q_i(Y_1,\ldots Y_n)}\right)^{\alpha} \mid T = \theta\right] \nu(d\theta)$$
$$= \int_{W^n} \left(\int_{\Theta_j} \prod_{m=1}^{n} l(w_m,\theta)\nu(d\theta) \right)^{\alpha} \times$$
$$\times \left(\int_{\Theta_i} \prod_{m=1}^{n} l(w_m,\theta)\nu(d\theta) \right)^{1-\alpha} d\sigma^n$$
$$= \left[\nu(\Theta_j)^{\alpha}\, \nu(\Theta_i)^{1-\alpha} \right]^{1-n} \times$$
$$\times \prod_{m=1}^{n} \int_W \left(\int_{\Theta_j} l(w,\theta)\nu(d\theta) \right)^{\alpha} \left(\int_{\Theta_i} l(w,\theta)\nu(d\theta) \right)^{1-\alpha} \sigma(dw)$$
$$= (q_{ij}^{\alpha})^{n-1} \int_W l_j(w)^{\alpha}\, l_i(w)^{1-\alpha}\, \sigma(dw)$$

where we put

$$l_j(w) := \int_{\Theta_j} l(w,\theta)\nu(d\theta) .$$

Now the assertion follows from noting that

$$P_\nu(T \notin \Theta_{T_n}) = \sum_{i \neq j} P_\nu(T = j\, , \, T \in \Theta_i)$$
$$\le q^{n-1} \inf_{0<\alpha<1} \int_W \sum_{i \neq j} l_j(w)^{\alpha} l_i(w)^{1-\alpha}\, \sigma(dw)$$

and

$$\inf_{0<\alpha<1} \int_W \sum_{i \neq j} l_j(w)^{\alpha}\, l_i(w)^{1-\alpha}\, \sigma(dw)$$
$$= \inf_{0<\alpha<1} \int_W \left[\sum_j l_j(w)^{\alpha} \sum_i l_i(w)^{1-\alpha} - \sum_m l_m(w) \right] \sigma(dw)$$

$$\leq \int_W \left[(\sum_i l_i(w)^{1/2})^2 - \sum_m l_m(w) \right] \sigma(dw)$$

$$\leq K - 1$$

where we used $\int_W \sum_m l_m(w)\ \sigma(dw) = 1$ and the Hölder inequality.

5. A Uniform Bound for the Regret

To apply Theorem 2 to our control model we select an $\tau_j \in \Theta_j$ for each j $1 \leq j \leq K$. Then we apply decision rule $f_{N-n}(\cdot, \tau_{T_n})$ on stage n, i.e. if $T_n = j$ we use one of the values $\theta \in \Theta_j$ for the adaptation of the decision.

THEOREM 3

Let (C) and (R_ε) hold for some $\varepsilon \geq 0$ given. Let (T_n) be chosen as in (6), $\tau_j \in \Theta_j$ $(1 \leq j \leq K)$ and assume that (7) holds.
Then for all $N \in \overline{\mathbb{N}}$, $n < N$ and all $\delta \in F^\infty$

$$E_{\nu,\delta}[\ V_T^{N-n}(X_n, Y_n) - V_T^{N-n}(X_n, Y_n, \delta^{N-n}(\tau_{T_n}))]$$
$$\leq 2\ (\rho_{N-n}\ \varepsilon + \bar{r}\ \frac{1-\beta^{N-n}}{1-\beta}\ (K-1)\ q^{n-1}\ )$$

where q is given by (8).
Note that the bound is again independent of the apriori distribution ν.

Proof : We write E instead of $E_{\nu,\delta}$. Put

$$U := V_T^{N-n}\ (X_n, Y_n) - V_T^{N-n}(X_n, Y_n, \delta^{N-n}(\tau_{T_n}))\ ,$$

then $0 \leq U \leq 2 \cdot \frac{1-\beta^{N-n}}{1-\beta}\ \bar{r}$. Put $\hat{\Theta}_n := \Theta_{T_n}$, then $T \in \hat{\Theta}_n$ implies $|T_n, T| \leq \varepsilon$, hence by Theorems 1 and 2

$$E\ U = E\ [E\ [U \mid T, T_n] 1_{[T \in \hat{\Theta}_n]}] + E\ U\ 1_{[T \notin \hat{\Theta}_n]}$$
$$\leq E\ [\ 2\rho_{N-n} |T, T_n| 1_{[T \in \hat{\Theta}_n]}] + 2\ \frac{1-\beta^{N-n}}{1-\beta}\ \bar{r}\ P(T \notin \hat{\Theta}_n)$$
$$\leq 2(\ \rho_{N-n}\ \varepsilon + 2\ \frac{1-\beta^{N-n}}{1-\beta}\ \bar{r}\ (K-1)\ q^{n-1}\ )$$

To obtain a small regret we can proceed as follows : first choose ε small such that (R_ε) holds, then choose a sample size such that the second term in the bound becomes small.

If $\Theta = \{\tau_1, \ldots \tau_K\}$ is finite we may choose $\varepsilon = 0$ in Theorems 2 and 3

With $|\theta,\theta'| := \begin{cases} 0 & \text{if } \theta=\theta' \\ 1 & \text{if } \vartheta=\theta' \end{cases}$ (R_0) holds for $\Theta_j := \{\tau_j\}$, $1\le j\le K$.

Then Theorem 2 coincides with a result in Rènyi (64) and the bound in Theorem 3 becomes

$$2\,\bar{r}\,\frac{1-\beta^{N-n}}{1-\beta}\,(K-1)\,q^{n-1} \le \frac{2\bar{r}(K-1)}{1-\beta}\,q^{n-1}$$

independently of (C), although (C) is easily fulfilled if Θ is finite.

If the Lipschitz - conditions (C) hold there is also another approach to find uniform bounds on the regret. In Kolonko (82) confidence intervals with fixed expected width are used. Instead of the error probability the significance level of the confidence interval leads to a bound on the effect of an estimation error. In that paper there are also given a number of examples for r_θ and μ_θ satisfying condition (C).

References

Rényi,A. (1964) On the Amount of Information Concerning an Unknown Parameter in a Sequence of Observations.
Publ.Math.Inst.Hung.Acad.Sci. 9,Ser. A, 617 -625 .

Vincze,I. (1967) On the Information-theoretical Foundation of Mathematical Statistics.
Proc.Coll. on Inf. Theory, Hungary, 1967, Ed.:A. Rényi, 503-509.

Kolonko,M (1982) Bounds for the Regret Loss in Dynamic Programming under Adaptive Control.
To appear in Zeitschrift für Operations Research.

STOCHASTIC REGRESSION MODELS AND CONSISTENCY OF THE LEAST SQUARES IDENTIFICATION SCHEME

Tze Leung Lai
Columbia University
New York, N.Y. 10027, USA

1. Introduction

Consider the multiple regression model

(1) $$y_n = \beta_1 x_{n1}+\dots+\beta_p x_{np}+\epsilon_n, \quad n=1,2,\dots,$$

where the ϵ_n are unobservable random errors, $\beta_1,\dots,\beta_p$ are unknown parameters, and y_n is the observed response corresponding to observed design levels $x_{n1},\dots,x_{np}$. When the design levels $x_{n1},\dots,x_{np}$ are random variables, the regression model (1) is called stochastic. Throughout the sequel we shall assume that the random errors ϵ_n form a martingale difference sequence with respect to an increasing sequence of σ-fields $\mathfrak{F}_n$, i.e., ϵ_n is $\mathfrak{F}_n$-measurable and $E(\epsilon_n|\mathfrak{F}_{n-1}) = 0$ a.s. Moreover, letting $\underset{\sim}{x}_n = (x_{n1},\dots,x_{np})'$, we shall assume that

(2) $\underset{\sim}{x}_n$ is $\mathfrak{F}_{n-1}$-measurable for all n.

The following examples of stochastic regression models have been extensively studied in the literature.

Example 1. The autoregressive model

(3) $$y_n = \beta_1 y_{n-1}+\dots+\beta_p y_{n-p}+\epsilon_n$$

in the time series literature is a special case of the stochastic regression model (1) with $\underset{\sim}{x}_n = (y_{n-1},\dots,y_{n-p})'$.

Example 2. An important stochastic model in the system and control literature is the input-output model

(4) $$y_n = \alpha_1 y_{n-1}+\dots+\alpha_k y_{n-k}+\gamma_1 u_{n-1}+\dots+\gamma_h u_{n-h}+\epsilon_n ,$$

where the errors ϵ_n form a martingale difference sequence, u_n is the input and y_n is the output at stage n. This is a special case of the stochastic regression model (1) with $\underset{\sim}{x}_n = (y_{n-1},\dots,y_{n-k},u_{n-1},\dots,u_{n-h})$

Example 3. Let $z_{ki}, w_j\ (i,j,k=1,2,\dots)$ be independent nonnegative integer-valued random variables such that z_{ki} have a common mean β and w_j have a common mean α. The Galton-Watson process

(5) $$y_n = \Sigma_{i=1}^{y_{n-1}} z_{ni} + w_n$$

can be written in the form

(6) $$y_n = \alpha+\beta y_{n-1}+\epsilon_n,$$

where $\epsilon_n = (w_n-\alpha) + \Sigma_{i=1}^{y_{n-1}}(z_{ni}-\beta)$. Note that $\{\epsilon_n,\mathfrak{F}_n,n\geq 1\}$ is a martingale difference sequence, where $\mathfrak{F}_n$ is the σ-field generated by $\{z_{ki}: i\leq y_{k-1}, k\leq n\}\cup\{w_k: k\leq n\}$. Thus, the Galton-Watson process is a special case of the stochastic regression model (1) with $\underset{\sim}{x}_n = (1,y_{n-1})'$.

Example 4. Suppose that in the regression model (1), the design levels $x_{n1},\ldots,x_{np}$ are chosen at stage n on the basis of previous observations $\underset{\sim}{x}_1,y_1,\ldots,\underset{\sim}{x}_{n-1},y_{n-1}$. Such adaptive choice of the design levels leads to a stochastic regression model. A well known example of adaptive designs is the Robbins-Monro stochastic approximation scheme for the simple linear model

(7) $$y_n = \alpha+\beta x_n+\epsilon_n .$$

The objective here is to choose the successive levels $x_1,x_2,\ldots$ in the model (7) so that the outputs $y_1,y_2,\ldots$ are as close as possible to a target value y^*. Assuming the sign of β to be known, say $\beta>0$, the Robbins-Monro stochastic approximation scheme chooses the design levels by the recursion

(8) $$x_{n+1} = x_n-c_n(y_n-y^*),$$

where $\{c_n\}$ is a sequence of positive constants such that

(9) $$\sum_1^\infty c_n^2 <\infty, \sum_1^\infty c_n = \infty .$$

For the regression model (1), let $X_n = (x_{ij})_{1\leq i\leq n, 1\leq j\leq p}$ denote the design matrix at stage n and let $Y_n = (y_1,\ldots,y_n)'$. The least squares estimate at stage n of the unknown parameter vector $\underset{\sim}{\beta} = (\beta_1,\ldots,\beta_p)'$ is given by

(10) $$\underset{\sim}{b}_n = (b_{n1},\ldots,b_{np})' = (X_n'X_n)^{-1}X_n'Y_n ,$$

where the inverse denotes the Moore-Penrose generalized inverse. In this paper we review some recent results on the question of strong consistency of $\underset{\sim}{b}_n$ in stochastic regression models. Section 2 considers the general stochastic regression model, and in Sections 3,4 and 5, we discuss the application of these results to the particular models in Examples 1,2 and 4. For the Galton-Watson model of Example 3, strong consistency of the least squares estimates has been established in the subcritical case $(\beta<1)$ by Heyde and Seneta [6].

2. Strong Consistency of Least Squares Estimates in Stochastic Regression Models

From (1) and (10), it follows that

$$(11) \qquad \underset{\sim}{b}_n - \underset{\sim}{\beta} = (X_n'X_n)^{-1} X_n(\epsilon_1,\ldots,\epsilon_n)' = (\sum_1^n \underset{\sim}{x}_i\underset{\sim}{x}_i')^{-1} \sum_1^n \underset{\sim}{x}_i\epsilon_i .$$

Throughout the sequel we shall assume that the martingale difference sequence $\{\epsilon_n, \mathfrak{F}_n, n\geq 1\}$ satisfies

$$(12) \qquad \sup_n E(\epsilon_n^2 | \mathfrak{F}_{n-1}) < \infty \text{ a.s.}$$

For the single-parameter $(p=1)$ case, we obtain from (11) and the martingale strong law that

$$b_n - \beta = \sum_1^n x_i\epsilon_i / \sum_1^n x_i^2 \to 0 \text{ a.s. on } \{\sum_1^\infty x_i^2 = \infty\}.$$

Therefore, in the case $p=1$,

$$(13) \qquad \sum_1^n x_i^2 \to \infty \text{ a.s.} \Rightarrow b_n \to \beta \text{ a.s.}$$

This result can be extended to the case of general p when the design levels $\underset{\sim}{x}_i$ are independent of the errors ϵ_n.

Theorem 1. Suppose that in the regression model (1), the errors ϵ_n form a martingale difference sequence satisfying (12) and are independent of $\{\underset{\sim}{x}_i : i\geq 1\}$. Then

$$(14) \qquad \lambda_{\min}(X_n'X_n) \to \infty \text{ a.s.} \Rightarrow \underset{\sim}{b}_n \to \underset{\sim}{\beta} \text{ a.s.}$$

Here and in the sequel we use the notations $\lambda_{\max}(A)$ and $\lambda_{\min}(A)$ to denote the maximum and minimum eigenvalues of a symmetric matrix A. Theorem 1 was recently proved by Lai, Robbins and Wei [10] in a somewhat more general context.

The independence assumption between $\{\underset{\sim}{x}_i\}$ and $\{\epsilon_i\}$ in Theorem 1 is not satisfied in the stochastic regression models of Examples 1-4. The following counter-example of Lai and Robbins [9], however, shows that (14) need not hold when the $\mathfrak{F}_{n-1}$-measurable design level $\underset{\sim}{x}_n$ is not independent of $\epsilon_1,\ldots,\epsilon_{n-1}$.

Example 5. Suppose that in the simple linear model (7) the errors $\epsilon_1, \epsilon_2, \ldots$ are i.i.d. with $E\epsilon_1 = 0$ and $E\epsilon_1^2 = 1$ and the design levels x_n satisfy the recursion

$$(15) \qquad x_{n+1} = \bar{x}_n + c\bar{\epsilon}_n \quad (x_1 = 0),$$

where $c\neq 0$ is a real constant and the notation $\bar{a}_n$ denotes the arithmetic mean of n numbers $a_1,\ldots,a_n$. Then the difference between the slope β and its least squares estimate $\hat{\beta}_n$ can be expressed as

$$\hat{\beta}_n-\beta = \sum_1^n(x_i-\bar{x}_n)\epsilon_i/\sum_1^n(x_i-\bar{x}_n)^2$$

$$= \{\sum_2^n(\tfrac{i-1}{i})(x_i-\bar{x}_{i-1})(\epsilon_i-\bar{\epsilon}_{i-1})\}/\{\sum_2^n(\tfrac{i-1}{i})(x_i-\bar{x}_{i-1})^2\},$$

and therefore by (15), $\hat{\beta}_n \to \beta-c^{-1}$ a.s. This in turn implies that $\hat{\alpha}_n-\alpha\to\sum_1^\infty i^{-1}\epsilon_i$ a.s., where $\hat{\alpha}_n$ is the least squares estimate of the intercept α. Moreover, in this case,

$$X_n'X_n = \begin{pmatrix} n & \sum_1^n x_i \\ \sum_1^n x_i & \sum_1^n x_i^2 \end{pmatrix},$$

and as shown in [11],

$$\lambda_{max}(X_n'X_n) \sim n\{1 + c^2(\sum_1^\infty i^{-1}\epsilon_i)^2\} \text{ a.s.},$$

$$\lambda_{min}(X_n'X_n) \sim c^2(\log n)/\{1 + c^2(\sum_1^\infty i^{-1}\epsilon_i)^2\} \text{ a.s.}$$

The above example shows that for stochastic regression models, the condition $\lambda_{min}(X_n'X_n) \to \infty$ a.s., or even the stronger condition $\liminf_{n\to\infty} \lambda_{min}(X_n'X_n)/\log \lambda_{max}(X_n'X_n) > 0$ a.s., is not sufficient for the strong consistency of $\underset{\sim}{b}_n$. Lai and Wei [11] recently obtained the following sufficient condition for the strong consistency of $\underset{\sim}{b}_n$ which, in view of the above example, is in some sense the weakest possible.

Theorem 2. Suppose that in the regression model (1), $\{\epsilon_n\}$ is a martingale difference sequence such that

(16) $\qquad \sup_n E(|\epsilon_n|^r|\mathcal{F}_{n-1}) <\infty$ a.s. for some $r>2$.

Assume that $\underset{\sim}{x}_n$ is $\mathcal{F}_{n-1}$-measurable for every n and that

(17) $\qquad \lambda_{min}(X_n'X_n)\to\infty$ and $\log \lambda_{max}(X_n'X_n) = o(\lambda_{min}(X_n'X_n))$ a.s.,

then $\underset{\sim}{b}_n \to \underset{\sim}{\beta}$ a.s.

Theorem 2 improves the recent result of Anderson and Taylor [1] who assumed the condition $\lambda_{max}(X_n'X_n) = O(\lambda_{min}(X_n'X_n))$ a.s., and of Christopeit and Helmes [3] who obtained the strong consistency of $\underset{\sim}{b}_n$ under the assumption $\lambda_{max}^\rho(X_n'X_n) = O(\lambda_{min}(X_n'X_n))$ a.s. for some $\rho>\frac{1}{2}$.

To apply Theorem 2 or the other aforementioned results in the literature, we have to estimate the order of magnitude of $\lambda_{max}(X_n'X_n)$ and of $\lambda_{min}(X_n'X_n)$. The former quantity can be easily estimated by the inequality

(18) $\qquad \lambda_{max}(X_n'X_n) \le tr(X_n'X_n) = \sum_{j=1}^p \sum_{i=1}^n x_{ij}^2 \le p\lambda_{max}(X_n'X_n)$.

The analysis of $\lambda_{min}(X_n'X_n)$, however, is often a much harder problem. A method of tackling this problem is discussed in [12]. This method is based on certain asymptotic properties of stochastic projections and the inequality

(19) $$p \min_{1\le j\le p} \|c_j(X_n)-\hat{c}_j(X_n)\|^2 \ge \lambda_{min}(X_n'X_n) \ge p^{-1} \min_{1\le j\le p} \|c_j(X_n)-\hat{c}_j(X_n)\|^2,$$

where $c_j(X_n)$ denotes the jth column vector of X_n, $\hat{c}_j(X_n)$ denotes the projection of $c_j(X_n)$ into the linear space spanned by the other column vectors of X_n, and $\|\underset{\sim}{u}\| = (\sum_1^n u_i^2)^{\frac{1}{2}}$ denotes the Euclidean norm of $\underset{\sim}{u} = (u_1,\ldots,u_n)'$.

While Theorem 2 considers the strong consistency of the entire vector $\underset{\sim}{b}_n$, the following theorem of Lai and Wei [12] gives analogous conditions (involving projections instead of eigenvalues) that would imply the strong consistency of b_{nj} for a particular j.

Theorem 3. Suppose that in the regression model (1), $\{\epsilon_n\}$ is a martingale difference sequence satisfying (16) and that $\underset{\sim}{x}_n$ is $\mathfrak{F}_{n-1}$-measurable for every n. For $j=1,\ldots,p$, the least squares estimate b_{nj} of β_j can be represented as

$$b_{nj} = \langle c_j(X_n)-\hat{c}_j(X_n),\ Y_n\rangle / \|c_j(X_n)-\hat{c}_j(X_n)\|^2 ,$$

where $\langle \underset{\sim}{u},\underset{\sim}{v}\rangle = \sum_1^n u_i v_i$ denotes the inner product of $\underset{\sim}{u}=(u_1,\ldots,u_n)'$ and $\underset{\sim}{v} = (v_1,\ldots,v_n)'$. If $\|c_j(X_n)-\hat{c}_j(X_n)\|\to\infty$ a.s. and

(20) $$\max_{k\ne j} \log^+\left(\sum_{i=1}^n x_{ik}^2\right) = o(\|c_j(X_n)-\hat{c}_j(X_n)\|^2) \text{ a.s.},$$

then $b_{nj} \to \beta_j$ a.s.

From (18) and (19), it follows that $\max_{1\le k\le p} \sum_{i=1}^n x_{ik}^2 \le p \max(X_n'X_n)$ and that $\lambda_{min}(X_n'X_n) \le p \min_{1\le j\le p} \|c_j(X_n)-\hat{c}_j(X_n)\|^2$. Hence, if condition (17) of Theorem 2 is satisfied, then condition (20) of Theorem 3 is also satisfied for every j. Therefore, Theorem 2 can be obtained as a corollary of Theorem 3.

3. Application to Autoregressive Models

Consider the autoregressive AR(p) model (3) of Example 1, where the random errors ϵ_n are assumed to form a martingale difference sequence such that (16) holds and $\liminf_{n\to\infty} E(\epsilon_n^2|\mathfrak{F}_{n-1})>0$ a.s. A commonly used estimate of the parameter vector $\underset{\sim}{\beta} = (\beta_1,\ldots,\beta_p)'$ is the least squares estimate

$$\underset{\sim}{b}_n = (X_n'X_n)^{-1} X_n'(y_{p+1},\ldots,y_n)' , \quad n>p,$$

where

$$X_n = \begin{pmatrix} y_p \cdots y_1 \\ y_{p+1} \cdots y_2 \\ y_{n-1} \cdots y_{n-p} \end{pmatrix} .$$

As shown by Lai and Wei [13],

(21) $$\liminf_{n\to\infty} n^{-1} \lambda_{min}(X_n'X_n)>0 \quad \text{a.s.}$$

Let $\varphi(z) = z^p-\beta_1 z^{p-1}-\ldots-\beta_p$ denote the characteristic polynomial of the AR(p) model. If all the roots z_j of $\varphi(z)$ lie on or inside the unit circle, i.e., $|z_j|\le 1$ for $j=1,\ldots,p$, then $y_n=o(n^{\frac{1}{2}})$ a.s. if $\max_{1\le j\le p} |z_j|<1$ and $y_n=o(n^{\rho-\frac{1}{2}}(\log\log n)^{\frac{1}{2}})$ a.s. otherwise, where ρ is the largest multiplicity of all the distinct roots of $\varphi(z)$ on the unit circle (cf. [13]). Hence $\lambda_{max}(X_n'X_n) = o(n^a)$ a.s. for some $a>1$, and therefore in view of (21), $\log \lambda_{max}(X_n'X_n) = o(\lambda_{min}(X_n'X_n))$ a.s. Theorem 2 is then applicable and gives the strong consistency of $\underset{\sim}{b}_n$.

Suppose all roots of $\varphi(z)$ lie outside the unit circle. Let

$$B = \begin{pmatrix} \beta_1 \cdots \beta_{p-1} & \beta_p \\ I_{p-1} & 0 \end{pmatrix}$$

denote the associated companion matrix, where I_k denotes the kxk identity matrix. Then in view of (11),

(22) $$\underset{\sim}{b}_n - \underset{\sim}{\beta} = (X_n'X_n)^{-1} \sum_{i=p+1}^{n} \underset{\sim}{x}_i \epsilon_i ,$$

where $\underset{\sim}{x}_n = (y_{n-1},\ldots,y_{n-p})'$. Since the modulus of each eigenvalue of B exceeds 1, it can be shown that with probability 1

(23) $$\lim_{n\to\infty} B^{-n} X_n'X_n(B^{-n})' \quad \text{exists and is positive definite, and}$$

(24) $$\lim_{n\to\infty} \sum_{i=p+1}^{n} \|B^{-n} \underset{\sim}{x}_i\| \quad \text{exists and is finite,}$$

(cf. [13]). From (22), it follows that

(25) $$\|\underset{\sim}{b}_n - \underset{\sim}{\beta}\|\le\|(B^{-n})'\|\|(B^{-n} X_n'X_n(B^{-n})')^{-1}\|(\sum_{i=1}^{n} \|B^{-n} \underset{\sim}{x}_i\|)\max_{1\le i\le n} |\epsilon_i|.$$

Since $\max_{1\le i\le n} |\epsilon_i| = o(n^{\frac{1}{2}})$ a.s. and $\log\|(B^{-n})'\| \sim -n\log(\min_{1\le j\le p} |z_j|)$, the strong consistency of $\underset{\sim}{b}_n$ follows from (23), (24) and (25).

In the general case where some roots of $\varphi(z)$ lie outside the unit circle and the remaining roots lie on or inside the unit circle, Lai and Wei [13] recently established the strong consistency of $\underset{\sim}{b}_n$

by using a linear transformation to decompose $\{y_n\}$ into two autoregressive processes $\{u_n\}$ and $\{v_n\}$, so that the characteristic polynomial of $\{u_n\}$ has all its roots on or inside the unit circle while the characteristic polynomial of $\{v_n\}$ has all its roots outside the unit circle.

4. Application to Identification and Control of Dynamic Systems

Consider the input-output system (4) of Example 2. In engineering applications, the least squares estimate $\underset{\sim}{b}_n$ of the parameter vector $\underset{\sim}{\beta} = (\alpha_1,\ldots,\alpha_k,\gamma_1,\ldots,\gamma_h)'$ is often used in its recursive form of the Kalman filter type (cf. [5]):

$$\underset{\sim}{b}_{n+1} = \underset{\sim}{b}_n + \{(y_{n+1} - \underset{\sim}{x}'_{n+1}\underset{\sim}{b}_n)/(1+\underset{\sim}{x}'_{n+1}V_n\underset{\sim}{x}_{n+1})\}V_n\underset{\sim}{x}_{n+1},$$

$$V_{n+1} = V_n - V_n\underset{\sim}{x}_{n+1}\underset{\sim}{x}'_{n+1}V_n/(1 + \underset{\sim}{x}'_{n+1}V_n\underset{\sim}{x}_{n+1}) .$$

This recursive representation of $\underset{\sim}{b}_n$ provides a simple algorithm for the recursive on-line identification of the system. In adaptive control systems, the sequentially updated least squares estimate $\underset{\sim}{b}_n$ is often used to decide on the input at the next stage. The underlying idea is that in many stochastic control problems, the optimal controller has a simple recursive form when the parameter $\underset{\sim}{\beta}$ of the system is known (cf. [2], [5]). Replacing $\underset{\sim}{\beta}$ by the least squares estimate $\underset{\sim}{b}_n$ in the optimal controller at each stage n, one may hope that the performance of such an adaptive controller approaches that of the optimal controller assuming known $\underset{\sim}{\beta}$ if $\underset{\sim}{b}_n$ should converge to $\underset{\sim}{\beta}$ with probability 1. Therefore there has been considerable interest in the question of strong consistency of $\underset{\sim}{b}_n$ in the engineering literature (cf. [5], [14], [15], [16]).

Theorem 2 enables us to improve earlier results in the literature on the consistency of the least squares identification scheme for the dynamic system (4). In particular, for non-explosive systems in which the roots of the characteristic polynomial $\varphi(z) = z^k - \alpha_1 z^{k-1} - \cdots - \alpha_k$ lie on or inside the unit circle and the inputs u_n satisfy the condition $u_n = O(n^d)$ a.s. for some $d>0$, Theorem 2 implies that the least squares identification scheme is strongly consistent if

$$\lim_{n\to\infty} \lambda_{\min}\Big(\sum_1^n \underset{\sim}{x}_i\underset{\sim}{x}'_i\Big)/\log n = \infty \text{ a.s.} \tag{26}$$

(cf. [11]), where $\underset{\sim}{x}_n = (y_{n-1},\ldots,y_{n-k},u_{n-1},\ldots,u_{n-h})'$.

5. Application to Adaptive Stochastic Approximation Schemes

Consider the Robbins-Monro stochastic approximation scheme (8) for the linear model (7) with i.i.d. errors ϵ_n such that $E\epsilon_1=0$ and $0<E\epsilon_1^2 = \sigma^2<\infty$. It is known (cf. [4]) that an asymptotically optimal choice of c_n for the stochastic approximation scheme (8) is $c_n = (n\beta)^{-1}$. In this case, letting $\theta = (y^*-\alpha)/\beta$, we have

(27) $x_n \to \theta$ a.s., and in fact,

$$\limsup_{n\to\infty} (n/2 \log\log n)^{\frac{1}{2}} |x_n - \theta| = \sigma/\beta \text{ a.s.,}$$

(28) $$n^{\frac{1}{2}}(x_n - \theta) \xrightarrow{\mathcal{D}} N(0, \sigma^2/\beta^2),$$

(29) $$\sum_1^n (x_i - \theta)^2 \sim (\sigma^2/\beta^2)\log n,$$

(cf. [7]). In ignorance of β, it is natural to try using $c_n = (nb_n)^{-1}$ in (8), where $b_n = b_n(x_1, y_1, \ldots, x_n, y_n)$ is some estimate of β based on the data already observed. Lai and Robbins [7, Theorem 3] showed that if b_n is strongly consistent, then the asymptotic properties (27), (28), (29) also hold for the adaptive stochastic approximation scheme

(30) $$x_{n+1} = x_n - (y_n - y^*)/(nb_n).$$

An obvious choice for b_n in the stochastic approximation scheme (30) is the usual least squares estimate

(31) $$\hat{\beta}_n = \left[\sum_1^n (x_i - \bar{x}_n) y_i\right] / \sum_1^n (x_i - \bar{x}_n)^2 .$$

Theorem 3 says that a sufficient condition for $\hat{\beta}_n$ to converge a.s. to β is that

(32) $$\sum_1^n (x_i - \bar{x}_n)^2 / \log n \to \infty \text{ a.s.}$$

However, if $\hat{\beta}_n \to \beta$ a.s., then (27), (28) and (29) would hold for the adaptive stochastic approximation scheme (30) with $b_n = \hat{\beta}_n$, so

$$\sum_1^n (x_i - \bar{x}_n)^2 = \sum_1^n (x_i - \theta)^2 - n(\bar{x}_n - \theta)^2 \sim (\sigma^2/\beta^2)\log n \quad \text{a.s.,}$$

by (27) and (29). Therefore the condition (32) cannot be satisfied by the scheme (30) if $b_n \to \beta$ a.s.

The relations (27) and (29) suggest another approach since they imply that with probability 1,

(33) $$\sum_1^n (x_i - \bar{x}_n)\epsilon_i = \sum_1^n (x_i - \theta)\epsilon_i - (\bar{x}_n - \theta)\sum_1^n \epsilon_i$$

$$= \sum_1^n (x_i - \theta)\epsilon_i + O(\log\log n),$$

and

(34) $$\sum_1^n (x_i - \bar{x}_n)^2 = \sum_1^n (x_i - \theta)^2 - n(\bar{x}_n - \theta)^2 \sim \sum_1^n (x_i - \theta)^2,$$

and therefore by (29) and the strong law for martingales,

$$\hat{\beta}_n = \{\sum_1^n (x_i-\bar{x}_n)\epsilon_i\}/\{\sum_1^n (x_i-\bar{x}_n)^2\} \to \beta \text{ a.s.}$$

Hence if $\hat{\beta}_n$ is indeed strongly consistent, then the sequence $\{x_n\}$ defined by (30) with $b_n = \hat{\beta}_n$ will behave nicely in the sense of (27) and (29), which in turn imply the strong consistency of $\hat{\beta}_n$ in view of (33) and (34). While this circular argument does not prove the strong consistency of $\hat{\beta}_n$ in adaptive stochastic approximation schemes, it suggests the usefulness of (33) and (34) in considering the strong consistency problem for $\hat{\beta}_n$. Combining (33) and (34) with the idea behind Theorem 3, Lai and Robbins [9] obtained the following

Theorem 4. Suppose that in the linear regression model (7), $\epsilon_1, \epsilon_2, \ldots$ are i.i.d. random variables with $E\epsilon_1 = 0$ and $E\epsilon_1^2 < \infty$, and the design level x_n at stage n is $\mathfrak{F}_{n-1}$-measurable for every n, where $\mathfrak{F}_n$ is the σ-field generated by $\epsilon_1, \ldots, \epsilon_n$. Assume that there exist constants $0<\gamma<\frac{1}{2}$ and θ such that with probability 1

(35) $\qquad x_n - \theta = o(n^{-\gamma})$, and

(36) $\qquad \liminf_{n\to\infty} \sum_1^n (x_i-\theta)^2/\log n > 0.$

Define $\hat{\beta}_n$ as in (33). Then $\hat{\beta}_n \to \beta$ a.s. on the event

$$\{\limsup_{n\to\infty} \sum_1^n (x_i-\theta)^2/\log n < \infty\} \cup \{\lim_{n\to\infty} \sum_1^n (x_i-\theta)^2/\log n = \infty\}.$$

In fact, there exists an event Ω_0 such that all sample points $\omega\in\Omega_0$ have the following property: Given $\delta>0$ and $\rho>0$, there exist positive numbers Δ, λ and N (depending on ω, δ, ρ) such that at ω, for all $n \ge N$,

(37a) $\qquad \sum_1^n (x_i-\theta)^2 \le \rho \log n \Rightarrow |\hat{\beta}_n - \beta| < \delta,$

(37b) $\qquad |\bar{x}_n - \theta| \le \lambda n^{-\frac{1}{2}}(\log n)^{\frac{1}{2}} \Rightarrow |\hat{\beta}_n - \beta| < \delta,$

(37c) $\qquad \sum_{i=1}^{[n^{2\gamma}]} (x_i-\theta)^2 \ge \Delta \log n \Rightarrow |\hat{\beta}_n - \beta| < \delta .$

In the case where positive bounds B and b are known such that $b<\beta<B$, as often occurs in practice, it is natural to truncate $\hat{\beta}_n$ above and below by B and b. For the adaptive stochastic approximation scheme $\{x_n\}$ defined by (30) with

(38) $\qquad b_n = b \vee (B \wedge \hat{\beta}_n),$

where $\vee$ and $\wedge$ denote maximum and minimum respectively, the fact that $b \le b_n \le B$ ensures that $\{x_n\}$ satisfies (35) and (36) with $\theta = (y^* - \alpha)/\beta$ (cf. [9]). Moreover, it also implies that there exists an event Ω_1 such that $P(\Omega_1) = 1$ and all sample points $\omega\in\Omega_1$ have the following property: There exist positive constants C,M,k (depending on ω) such that at ω, for all $\ell \geq M$ and $m \geq \ell^k$,

(39) $\quad 3\beta/2 \geq b_n (\geq b)$ for all $\ell \leq n \leq m$

$$\Rightarrow |\bar{x}_n - \theta| \leq Cn^{-\frac{1}{2}}(\log\log n)^{\frac{1}{2}} \text{ for all } \ell^k \leq n \leq m + m^{\frac{1}{2}},$$

(cf. [8]).

To establish the strong consistency of $\hat{\beta}_n$ in the stochastic approximation scheme (30) with b_n defined by (38), we make use of Theorem 4 and (39). By Theorem 4, it suffices to show that $\hat{\beta}_n \to \beta$ a.s. on $D = \{\limsup_{n\to\infty} \sum_1^n (x_i - \theta)^2/\log n = \infty\}$. Let $\delta > 0$ such that $b < \beta - \delta$ and $\beta + \delta < \min\{3\beta/2, B\}$, and let $\omega \in D \cap \Omega_0 \cap \Omega_1$. Since $\omega \in D$, we can choose an integer L large enough such that at ω, for $L \leq n \leq L^k$,

$$\Sigma_{i=1}^{[n^{2\gamma}]} (x_i - \theta)^2 > \Sigma_{i=1}^{[L^{2\gamma}]} (x_i - \theta)^2 > k\Delta \log L \geq \Delta \log n.$$

Hence by (37c), $|\hat{\beta}_n - \beta| < \delta$ at ω for $L \leq n \leq L^k$. In view of (39), this in turn implies that for $\nu = L^k + 1$, $|\bar{x}_\nu - \theta| \leq C\nu^{-\frac{1}{2}}(\log\log \nu)^{\frac{1}{2}}$ at ω, and therefore $|\hat{\beta}_\nu - \beta| < \delta$ at ω by (37b). Proceeding inductively in this way, we then obtain that at ω, for $j = 1, 2, \ldots$,

$$|\hat{\beta}_n - \beta| < \delta \text{ for } L \leq n \leq L^k + j$$
$$\Rightarrow |\bar{x}_n - \theta| \leq Cn^{-\frac{1}{2}}(\log\log n)^{\frac{1}{2}} \text{ for } n = L^k + j + 1, \text{ by (39)},$$
$$\Rightarrow |\hat{\beta}_n - \beta| < \delta \text{ for } n = L^k + j + 1, \text{ by (37b)}.$$

Hence $|\hat{\beta}_n - \beta| < \delta$ at ω for all $n \geq L$. Since δ is arbitrary, $\hat{\beta}_n \to \beta$ at ω.

In the case where prior upper and lower bounds for β are not known, Lai and Robbins [9] introduced a stochastic truncation of $\hat{\beta}_n$ in place of (38) and extended the above argument to this more general case. In addition, the argument also works for non-linear regression models of the form $y_n = M(x_n) + \epsilon_n$, where the regression function $M(x)$ satisfies the following conditions:

(40a) $\quad M(\theta) = y^*$ and $M'(\theta) = \beta$ exists and is positive,

(40b) $\quad \inf_{\delta \leq |x-\theta| \leq \delta^{-1}} \{(M(x) - y^*)(x - \theta)\} > 0$ for every $0 < \delta < 1$,

(40c) $\quad |M(x)| \leq A|x| + D$ for some $A, D > 0$ and all x.

References

1. ANDERSON, T.W. and TAYLOR, J. (1979). Strong consistency of least squares estimators in dynamic models. Ann. Statist. 7, 484-489.

2. ASTRÖM, K.J. (1970). Introduction to Stochastic Control Theory. Academic Press, New York.

3. CHRISTOPEIT, N. and HELMES, K. (1980). Strong consistency of least squares estimators in linear regression models. Ann. Statist. 8, 778-788.

4. CHUNG, K.L. (1954). On a stochastic approximation method. Ann. Math. Statist. 25, 463-483.

5. GOODWIN, G.C. and PAYNE, R.L. (1977). Dynamic System Identification. Academic Press, New York.

6. HEYDE, C.C. and SENETA, E. (1972). Estimation theory for growth and immigration rates in a multiplicative process. J. Appl. Probability 9, 235-256.

7. LAI, T.L. and ROBBINS, H. (1979). Adaptive design and stochastic approximation. Ann. Statist. 7, 1196-1221.

8. LAI, T.L. and ROBBINS, H. (1979). Local convergence theorems for adaptive stochastic approximation schemes. Proc. Nat. Acad. Sci. U.S.A. 74, 2667-2669.

9. LAI, T.L. and ROBBINS, H. (1981). Consistency and asymptotic efficiency of slope estimates in stochastic approximation schemes. Z. Wahrscheinlichkeitstheorie verw. Gebiete 56, 329-360.

10. LAI, T.L. ROBBINS, H. and WEI, C.Z. (1979). Strong consistency of least squares estimates in multiple regression II. J. Multivariate Anal. 9, 343-361.

11. LAI, T.L. and WEI, C.Z. (1982). Least squares estimates in stochastic regression models with applications to identification and control of dynamic systems. Ann. Statist. 10, 154-166.

12. LAI, T.L. and WEI, C.Z. (1982). Asymptotic properties of projections with applications to stochastic regression problems. To appear in J. Multivariate Anal.

13. LAI, T.L. and WEI, C.Z. (1983). Asymptotic properties of general autoregressive models and strong consistency of least squares estimates of their parameters. To appear in J. Multivariate Anal.

14. LJUNG, L. (1976). Consistency of the least squares identification method. IEEE Trans. Automat. Contr. AC-21, 779-781.

15. LJUNG, L. (1977). Analysis of recursive stochastic algorithms. IEEE Trans. Automat. Contr. AC-22, 551-575.

16. MOORE, J.B. (1978). On strong consistency of least squares identification algorithms. Automatica 14, 505-509.

RECURSIVE IDENTIFICATION TECHNIQUES

Lennart Ljung
Division of Automatic Control
Dept. of Electrical Engineering, Linköping University
S-581 83 Linköping, Sweden

ABSTRACT

Some basic results on recursive identification techniques and their properties are reviewed. The link between adaptive algorithms, recursive identification, and off-line identification is stressed. The fundamental character of the prediction and its gradient with respect to the adjustable parameters is pointed out.

1. INTRODUCTION

A fundamental problem in modern science is that of building mathematical models of various real-life objects and systems. The term system identification is commonly used for the problem of constructing mathematical models of dynamical systems based on input-output signals. In some cases it is necessary to construct the models on-line, at the same time as the data is sequentially collected from the system. This problem is handled by recursive identification techniques. The same problem is known as sequential parameter estimation in the statistical literature and adaptive (or adaptation) algorithms in the signal processing field.

In this contribution we shall highlight some basic ideas of recursive identification techniques, as well as to stress the close relationship to the general system identification problem. The presentation is by necessity a brief one. Further details on the topic can be found in Ljung and Söderström (1982) and Ljung (1979, 1981).

By recursive identification we mean the following: A set of candidate models $\mathcal{M}$ is given. The set is parametrized by a finite dimensional parameter vector θ, such that a given model in the set is denoted by $\mathcal{M}(\theta)$. Also, a sequence of measurements $z(t)$; $t=1,2,\ldots$ from the system in question is given, such that $z(t)$ is the measurement vector obtained at time t. The problem is to, at time t, determine a model $\mathcal{M}(\hat{\theta}(t))$ in the set $\mathcal{M}$ that describes the data $z(j)$; $j=1,\ldots,t$ well, in some sense. The calculation of $\hat{\theta}(t)$ is constrained to be recursive, i.e. it should be calculated using only a fixed, and t-independent amount of memory space and computations. Conceptually we could write

$$\begin{aligned} S(t) &= F(t,S(t-1),z(t)) \\ \hat{\theta}(t) &= H(S(t)) \end{aligned} \tag{1}$$

where S(t) is a finite dimensional auxiliary vector ("an information state"). Usually, as we shall see, the algorithm takes a more specific form.

Now, several different approaches to the construction of such recursive algorithms can be taken. Some common ones are

- <u>The Bayesian approach</u>: The model parameter θ is considered to be a random vector, which of course is correlated with the data process $\{z(t)\}$. (Approximations of) the a posteriori probability density for θ can then be computed, and $\hat{\theta}(t)$ can be taken as, e.g., the mean of this posterior distribution.

- <u>The Stochastic Approximation approach</u>: A measure of fit between the data and the model is chosen, e.g., the variance of the difference of true output and model output. This criterion can then be minimized using ideas from stochastic approximation, like the Robbins and Monro (1951) algorithm.

- <u>Observers, Model-reference techniques, Pseudolinear regressions</u>: The model error can be correlated with part of the data in order to make it as small as possible.

In this contribution, though, we shall concentrate on a fourth approach, viz to derive recursive algorithms from off-line identification ideas.

2. MODELS

Many different model sets can be used for a given system. It is the task of the user to find a suitable model set that matches the purpose of the modelling. It will be convenient to single out one part of the measurement vector $z(t)$ as as "output" $y(t)$ (which could be equal to $z(t)$) and term the rest an "input" $u(t)$:

$$z^T(t) = (y^T(t)u^T(t)). \tag{2}$$

A typical feature of any model $M(\theta)$ is that it provides a prediction or "guess" of the next output, given previous measurements:

$$\hat{y}(t|\theta) = g_M(\theta;t,z^{t-1}) \tag{3}$$

$$z^{t-1} = (z(t-1)\dots z(t))$$

Here $g_M(\theta;t,\cdot)$ is a function from R^N to R^P where $p=\dim y$ and $N=(t-1)\cdot\dim z$. In a common family of models the prediction is computed by linear, finite dimensional filtering:

$$\begin{aligned}\varphi(t+1,\theta) &= F(\theta)\varphi(t,\theta) + G(\theta)z(t)\\ \hat{y}(t|\theta) &= H(\theta)\varphi(t,\theta)\end{aligned} \tag{4}$$

As a simple example we may take a linear regression model

$$\hat{y}(t|\theta) = -a_1y(t-1)-\dots-a_ny(t-n)+b_1u(t-1)+\dots+b_mu(t-m) \tag{5}$$

where

$$\theta^T = (a_1\dots a_n\ b_1\dots b_m). \tag{6}$$

With the notation

$$\varphi^T(t)=(-y(t-1)\dots-y(t-n)\ u(t-1)\dots u(t-m)) \tag{7}$$

we could rewrite (5) in the following convenient form

$$y(t|\theta) = \varphi^T\ (t). \tag{8}$$

The general descriptions (3) and (4) include models that are much more sophisticated than (8). Let us here only mention an ARMA model of a signal y(t):

$$y(t) + a\ y(t-1) = e(t) + c\ e(t-1) \tag{9}$$

Here $\{e(t)\}$ is white noise, and we have confined ourselves to a first order model for simplicity.

It is fairly starightforward to show that with $\theta^T = (a\ c)$, the predictor corresponding to (9) is given by

$$\hat{y}(t|\theta) + c\ \hat{y}(t-1|\theta) = (c-a)\ y(t-1). \tag{10}$$

For a more detailed exposé over different model sets and how they relate to the general description (3) we may refer to Ljung and Söderström (1983).

3. IDENTIFICATION CRITERIA

From the predicted output $\hat{y}(t|\theta)$ according to the model $M(\theta)$ and the recorded output $y(t)$ we can compute the prediction error

$$\varepsilon(t,\theta) = y(t) - \hat{y}(t|\theta). \tag{11}$$

The identification problem is to determine a suitable estimate $\hat{\theta}_N$, based on z^N. Almost all identification methods are based on the following idea.

"Select $\hat{\theta}_N$, so that the sequence
$\varepsilon(t,\hat{\theta}_N)$ $t=1,\ldots,N$ becomes 'small'". (12)

Two common ways of qualifying what should be meant by "small" are the following ones:

<u>Criterion minimization</u>:

Select $\hat{\theta}_N$ so that the criterion

$$V_N(\theta,z^N) = \frac{1}{N}\sum_{t=1}^{N} |\varepsilon(t,\theta)|^2 \tag{13}$$

is minimized. For the model (8) this approach gives the well known least squares method.

Of course, other "norms", than the quadratic one in (13) can be chosen. A general criterion would be

$$V_N(\theta,z^N) = \frac{1}{N}\sum_{t=1}^{N} \ell(t,\theta,\varepsilon(t,\theta)) \tag{14}$$

where $\ell(t,\theta,\varepsilon)$ is a suitably chosen scalar-valued function. Within the family (14) we find several well known methods, such as the maximum likelihood method, corresponding to

$$\ell(t,\theta,\varepsilon) = -\log f(t,\theta,\varepsilon),$$

where f is the probability density function for the prediction error $\varepsilon(t,\theta)$. Other examples of methods within the family (14) are robust methods, output error methods, etc.

<u>Correlation approach</u>:

Select a sequence of vectors

$\zeta(t,\theta),\ t=1\ \ldots,\ N,$

$\zeta(t,\theta) = h(z^{t-1},\theta)$

representing relevant information about the system available at time t-1. Then choose $\hat{\theta}_N$ so that $\{\varepsilon(t,\hat{\theta}_N)\}$ is uncorrelated with $\{\zeta(t,\hat{\theta}_N)\}$:

$$\frac{1}{N}\sum_{1}^{N} \zeta(t,\theta)\varepsilon(t,\theta) = 0 \Rightarrow \theta=\hat{\theta}_N \tag{15}$$

For the model (8) this gives the wellknown instrumental variable method. Then $\zeta(t,\theta)$ is often chosen as a function of past inputs, independent of θ. Within the family (15) we also find pseudolinear regressions and model reference techniques.

Now, for a practical identification method we also need a numerical algorithm to actually minimize (13) or solve (15). We shall now briefly discuss some general methods for the first problem.

Except in the case (8), where (13) is quadratic in θ, the minimization is performed by iterative techniques. With i as the iteration number we proceed as follows:

$$\hat{\theta}_N^{(i)} = \hat{\theta}_N^{(i-1)} - \mu_N^{(i)}\,[R_N^{(i)}]^{-1}\,V'_N(\hat{\theta}_N^{(i-1)},\ z^N) \tag{16}$$

Here V'_N is the gradient of the criterion

$$V'_N(\theta,z^N) = -\frac{1}{N}\sum_{t=1}^{N} \psi(t,\theta)\varepsilon(t,\theta) \tag{17}$$

where we defined

$$\psi(t,\theta) = \frac{d}{d\theta}\hat{y}(t|\theta). \tag{18}$$

R_N is a matrix that possibly modifies the search direction from the gradient one. Typical choices are:

$$R_N^{(i)} = I \text{ or } R_N^{(i)} = |V''_N(\theta_N^{(i-1)},z^N|^2 \cdot I \tag{19}$$

(gradient or steepent descent, un-normalized and normalized)

$$R_N^{(i)} = V_N''(\hat{\theta}_N^{(i-1)}, z^N) \quad \text{(Newton)} \tag{20}$$

$$R_N^{(i)} = \frac{1}{N}\sum_{1}^{N} \psi(t,\hat{\theta}_N^{(i-1)})\psi^T(t,\hat{\theta}_N^{(i-1)}) \quad \text{(Gauss-Newton)} \tag{21}$$

The scalar μ is chosen so as to guarantee a decrease in the criterion value.

Similar approaches apply also to the solution of (15).

In addition to these general techniques, special schemes for special model sets have also been devised that utilize some particular structures of $\varepsilon(t,\theta)$ and $\psi(t,\theta)$.

For specific model sets the iterative technique (16), (21) need only be complemented by schemes for computation of $\hat{y}(t|\theta)$ and $\psi(t,\theta)$ for any given value of θ. The general linear model (4) gives, conceptually, a linear filter for this:

$$\xi(t+1,\theta) = A(\theta)\xi(t,\theta) + B(\theta)z(t)$$

$$\begin{pmatrix} \hat{y}(t|\theta) \\ \mathrm{col}\,\psi(t,\theta) \end{pmatrix} = C(\theta)\xi(t,\theta) \tag{22}$$

Here "col" means that the columns of ψ are stacked on top of each other.

4. RECURSIVE IDENTIFICATION

It is clear that the general scheme (16) does in general not comply with the structure (1) for a recursive algorithm. However, we can modify (16) to achieve this. Conceptually, this is obtained by making a one additional iteration i-1→i at the same time as the data record is advanced from N-1 to N. From (16) we thus obtain

$$\hat{\theta}(t) = \hat{\theta}(t-1)-\mu(t)R^{-1}(t)V_V'(\hat{\theta}(t-1),z^t)$$

where we introduced the notation

$$\hat{\theta}(t) = \hat{\theta}_t^{(t)} \text{ etc.}$$

Now, from (17) we see that

$$V_t'(\theta,z^t) = \frac{t-1}{t} V_{t-1}'(\theta,z^{t-1}) - \frac{1}{t} \psi(t,\theta)\varepsilon(t,\theta) \tag{23}$$

Introducing the approximation that we assume $\hat{\theta}(t-1)$ that actually minimized $V_{t-1}(\theta,z^{t-1})$, i.e.

$$V_{t-1}'(\hat{\theta}(t-1),z^{t-1}) = 0 \tag{24}$$

now gives the algorithm

$$\hat{\theta}(t)=\hat{\theta}(t-1)+\gamma(t)R^{-1}(t)\psi(t,\hat{\theta}(t-1))\varepsilon(t,\hat{\theta}(t-1)) \tag{25}$$

where we introduced

$$\gamma(t) = \mu(t)/t \tag{26}$$

In case $\psi(t,\theta)$ and $\varepsilon(t,\theta)$ can be computed for any θ using only a fixed amount (t-independent) of past data, then (5) is indeed a recursive algorithm. This is the case for the model (8), where

$$\begin{aligned} \varepsilon(t,\theta) &= y(t) - \theta^T\varphi(t) \\ \psi(t,\theta) &= \varphi(t). \end{aligned} \tag{27}$$

With $\mu=1$,

$$R(t) = \frac{1}{t} \sum_{k=1}^{t} \varphi(k)\varphi^T(k) \tag{28}$$

(the Newton or the Gauss-Newton choice), (25)-(28) is the celebrated recursive least squares algoirthm. For the gradient choice (19) we obtain the LMS-algorithm of Widrow, well known from many signal processing applications, e.g. Widrow et al (1975).

In general, it is not possible to compute $\varepsilon(t,\hat{\theta}(t-1))$ and $\psi(t,\hat{\theta}(t-1))$ using a fixed amount of data. In such a case these quantities have to be replaced by recursively computed quantities, denoted by $\varepsilon(t)$ and $\psi(t)$. The following simple example shows how to do this:

Consider the model (9). From (10) we easily find how to compute $\varepsilon(t,\theta)$ and $\psi(t,\theta)$ for any θ:

$$\varepsilon(t,\theta) + c\ \varepsilon(t-1,\theta) = y(t) + a\ y(t-1) \tag{29}$$

$$\psi(t,\theta) + c\ \psi(t-1,\theta) = \begin{pmatrix} -y(t-1) \\ \\ \varepsilon(t-1,\theta) \end{pmatrix} \tag{30}$$

Now, the obvious way of computing the approximations $\varepsilon(t)$ and $\psi(t)$ is to replace the nominal value θ in (29) - (30) by the current estimate $\hat{\theta}(t-1)$:

$$\varepsilon(t) + \hat{c}(t-1)\varepsilon(t-1) = y(t) + \hat{a}(t-1)y(t-1) \tag{31}$$

$$\psi(t) + \hat{c}(t-1)\psi(t-1) = \begin{pmatrix} -y(t-1) \\ \varepsilon(t-1) \end{pmatrix} \tag{32}$$

The complete algorithm for estimating a and c in the ARMA model (9) thus is (31), (32) together with:

$$\hat{\theta}(t) = \hat{\theta}(t-1) + \gamma(t)R^{-1}(t)\psi(t)\varepsilon(t) \tag{33}$$

For the Gauss-Newton algorithm we take

$$R(t) = R(t-1) + \frac{1}{t}\left[\psi(t)\psi^T(t) - R(t-1)\right] \tag{34}$$

Obviously, the same idea can be applied to any model set (3):

"Derive the equations defining $\varepsilon(t,\theta)$ and $\psi(t,\theta)$ for any nominal value θ. In order to obtain the approximations $\varepsilon(t)$ and $\psi(t)$ replace at each time instant the nominal value in these expressions by the latest available estimate".

Applied to the linear model (4) (see also (22)) this gives a general recursive Gauss-Newton algorithm

$$\varepsilon(t) = y(t) - \hat{y}(t)$$

$$\hat{\theta}(t) = \hat{\theta}(t-1) + \gamma(t)R^{-1}(t)\psi(t)\varepsilon(t)$$

$$\xi(t+1) = A(\hat{\theta}(t))\xi(t) + B(\hat{\theta}(t))z(t) \tag{35}$$

$$\begin{pmatrix} \hat{y}(t+1) \\ \\ \mathrm{col}\ \psi(t+1) \end{pmatrix} = C(\hat{\theta}(t))\xi(t)$$

$$R(t) = R(t-1) + \gamma(t)[\psi(t)\psi^T(t)-R(t-1)]$$

For a general criterion function

$$\ell(\varepsilon(t,\theta))$$

as in (14) the factor $\psi(t)\varepsilon(t)$ should be replaced by $\psi(t)\ell'(\varepsilon(t))$.

The algorithm (35) can be applied to arbitrary model sets, yielding a large family of recursive identification methods. These are well known in special cases. We can also, for special models, use certain approximations for the gradient ψ to yield other families for methods, see e.g. Ljung and Söderström (1983) or Ljung (1981).

5. ASYMPTOTIC PROPERTIES OF THE RECURSIVE ESTIMATES

The recursively computed estimate $\hat{\theta}(t)$ is a fairly complicated function of z^t. Its statistical properties will thus be inherited from those of z^t in a complex manner. It is possible, though, to compute the asymptotic statistical properties of $\hat{\theta}(t)$ as t approaches infinity, for the general algorithm (35) applied to an arbitrary (linear) model set. The result can be quoted as follows:

"For any positive definite choice of R(t) in (35), $\hat{\theta}(t)$ converges with probability one to a (local) minimum of

$$V(\theta) = \lim_{N\to\infty} \frac{1}{N} \sum_{1}^{N} E|\varepsilon(t,\theta)|^2 \tag{36}$$

as t tends to infinity, if $\gamma(t)\to 0$".

"If θ^* is such a limit point where $\bar{V}(\theta^*)$ is non-singular, then, for the Gauss-Newton choice (35) $\sqrt{t}(\hat{\theta}(t)-\theta^*)$ AsN(0,P) supposing $t\cdot\gamma(t)\to 1$ as $t\to\infty$". (37)

(i.e. the right hand side converges in distribution to the normal distribution with zero mean and covariance matrix P.)

An explicit expression for P can be given. It is fairly complicated, though. It simplifies in the following special case.

"If θ^* is such that $\{\varepsilon(t,\theta^*)\}$ is a sequence of independent random variables with zero mean and covariance matrix I $P = [E\ \psi(t,\theta^*)\ \psi^T(t,\theta^*)]^{-1}$". (38)

We might add that the results (36)-(38) are the same that hold for the off-line estimate $\hat{\theta}_N$ defined by minimization of (13).

Notice that there is no assumption that the true system can be exactly described within the model set for (36)-(37) to hold. Result (36) thus states that $M(\hat{\theta}(t))$ converges to the best approximation of the system (if not caught in a non-global local minimum).

Notice also that (38) equals the Cramér-Rao lower bound for the covariance of any estimate, provided the conditions mentioned in the third result hold and the (t,) are gaussian. The recursive identification algorithm (33)-(35) is thus asymptotically efficient. Asymptotically, no other algorithm can perform better for the chosen model set.

Proofs for the statements in this section can be found in Ljung and Söderström (1983).

6. CONCLUSIONS

In this contribution we have discussed the close relationship between general recursive identification algorithm and the corresponding off-line identification problem. This in turn makes the connection between adaptive algorithms of various kinds and basic identification concepts. This connection is perhaps worth while to stress.

We have also shown the fundamental role of prediction for this problem.

REFERENCES

L Ljung, "Convergence of recursive estimators", Proc. 5th IFAC Symposium on Identification, Darmstadt, Sept. 1979. Survey paper S8, pp. 131-144.

L. Ljung,"Analysis of a general recursive prediction error algorithm", Automatica, Vol. 17, No. 1, pp. 89-99, 1981.

L. Ljung and T. Söderström,"Theory and Practice of Recursive Identification", MIT Press, Cambridge, MA, USA, 1983.

H. Robbins and S. Monro,"A stochastic approximation method", Ann. Math. Stat., Vol. 22, pp. 400-407, 1951.

B. Widrow et al,"Adaptive noise cancelling: Principles and Applications", Proc. IEEE, Vol. 63, pp. 1692-1716, 1975.

AN OPTIMIZATION PROBLEM FOR MATRICES WITH APPLICATION TO DECISION MODELS

Petr Mandl
Department of Probability and Mathematical Statistics, Charles University
Sokolovská 83, 186 00 Prague 8, Czechoslovakia

1. Introduction

The problem of choosing a matrix with maximal Perron-Frobenius eigenvalue and its extensions were studied in a series of papers ([4] - [5]). Recently, the connection between principal eigenvalues of operators and stochastic control has received attention in the works of W.H.Fleming, S.K.Mitter, I.Karatzas. They find a control problem with average cost criterion such that the minimal cost coincides with the principal eigenvalue.The control problem for the case when the operator is a quazi-nonnegative matrix can be found in [2]. In the present paper we shall show, how it can be used to construct algorithms for maximizing the principal eigenvalue of quazi-nonnegative matrices.

The basic model in our considerations will be a controlled Markov process $X = \{X_t, \ t \geq 0\}$ with finite state space I whose dynamics is defined by the transition rates

$$rq(i,j;z), \quad i \neq j, \quad i,j \in I.$$

z denotes the control parameter taking values in a set J assumed to be compact, and the rates continuous in z. r is a time scale factor supposed to be large $(r \to \infty)$. The optimization algorithms for such processes, particularly the Howard's iteration procedure , are widely known. The simpliest controls of process X are the stationary controls. They are represented by mappings u(i) from I to J. u(i) is the control parameter value employed, whenever X is in state i. Under the control u, X is a homogenous Markov process with transition rates

$$(1) \qquad rq(i,j;\, u(i)), \quad i \neq j, \quad i,j \in I.$$

The set of the stationary controls will be denoted by U.

<u>Example 1.</u> Reliability of a large system. Consider a system consisting of a large number of elements. Let the evolution of each element be a Markov process with transition rates (1), where u is

a stationary control. Further assume that each element in state i has failure rate

$$rd(i;u(i)) \geqq 0, \quad i \in I.$$

After failure the element ceases to operate. Let Y_t denote the number of operating elements at time t. The entire system fails, when this number reaches a given level a. The aim is to maximize by appropriate choice of u the probability that the system will operate at time T

$$(2) \qquad P(Y_T > a) = \max.$$

The approximate solution of (2) is presented in the next section.

2. Problem formulation

Let us make an assumption.

Assumption 1. For each $u \in U$ the matrix

$$(3) \qquad Q^u = \| q(i,j;\, u(i)) \|_{i,j \in I}$$

with

$$(4) \qquad q(i,i;u(i)) = - \sum_{j \neq i} q(i,j;\, u(i)) - d(i;\, u(i))$$

is indecomposable.

Q^u are matrices of the quazi-nonnegative type, for which a version of the Perron-Frobenius Theorem is valid. Namely, there exists a principal eigenvalue $-\lambda(u)$, which is real, and exceeds the real parts of other eigenvalues. The minus sign in the denotation is introduced for convenience. The eigenvectors corresponding to $-$ (u) are positive. Let $P^u(t)$, $t \geqq 0$, be the transition probability matrix satisfying

$$\frac{d}{dt} P^u(t) = r\, Q^u P^u(t), \qquad t \geqq 0.$$

From the above said follows

$$P^u(t) = A^u \exp\{-\lambda(u)rt\} + o(\exp\{-\lambda(u)rt\}), \quad r \longrightarrow \infty \quad .$$

Hence also

$$P(\text{ element operating at } t\ ) \asymp \exp\{-\lambda(u)\, rt\}.$$

We conclude that an approximate solution of (2) can be obtained by solving the subsequent problem.

Problem 1. Find $u \in U$ such that Q^u has maximal principal eigenvalue $-\lambda(u)$ or

(5) $\lambda(u) = \min.$

Next we explain the connection between eigenvalues and optimal control. Let Q be a matrix as in (3), $-\lambda$ its principal eigenvalue. Let $s = (s_i, i \in I)$ denote a positive vector. According to [2] λ fulfils

(6) $$\min_s \{ \sum_{j \neq i} q(i,j) [s_j(w(j) - w(i)) + 1 + s_j \log s_j - s_j] + d(i) - \lambda \} = 0, \quad i \in I,$$

for a set $w(j)$, $j \in I$, of auxiliary constants. To see this, evaluate the minimum in the square bracket. (6) then becomes

$$\sum_{j \neq i} q(i,j) [-\exp\{w(i) - w(j)\} + 1] + d(i) - \lambda = 0, \quad i \in I,$$

or, with regard to (4),

(7) $$\sum_j a(i,j) \exp\{-w(j)\} + \lambda \exp\{-w(i)\} = 0, \quad i \in I.$$

(7) says that λ is an eigenvalue with a positive eigenvector, and consequently the principal one.

To interpret (6) consider a controlled Markov process $V = \{V_t, t \geq 0\}$ with state space I and transition rates

$$q(i,j)s_j, \quad i \neq j, \quad i,j \in I.$$

s is the control parameter . Let there be cost arising from the process. The sojourn in state i implies the cost

(8) $$c(i; s) = \sum_{j \neq i} q(i,j)(1 + s_j \log s_j - s_j) + d(i)$$

per unit time. (6) then shows (see e.g. [3]) that λ is the minimal average cost per unit time attainable.

It is now straight-forward that Problem 1 means just the introduction of an additional control variable z into (6) and (8). Thus, we have for

$$\hat{\lambda} = \min_{u \in U} \lambda(u)$$

the equations

$$(9)\quad \min_{s,z}\Big\{c(i;z,s) + \sum_{j\neq i} q(i,j;z)s_j(w(j)-w(i)) - \hat{\lambda}\Big\} = 0, \qquad i\in I.$$

(9) can be solved using the Howard iteration method. The procedure is started by choosing a stationary control $({}^{o}u(i),\ {}^{o}s(i))$. At the n-th step, when $({}^{n}u(i),\ {}^{n}s(i))$ has already been defined, the system of equations

$$(10)\quad c(i;{}^{n}u(i),\ {}^{n}s(i)) + \sum_{j\neq i} q(i,j;\ {}^{n}u(i))\ {}^{n}s_j(i)({}^{n}w(i)-{}^{n}w(j)) - {}^{n}\lambda = 0, \qquad i\in I,$$

is solved for unknowns ${}^{n}\lambda$, ${}^{n}w_j$, $j\in I$. (10) determines ${}^{n}\lambda$ uniquely, ${}^{n}w_j$, $j\in I$, up to an additive constant. Next it is set

$${}^{n+1}u(i) = \hat{z}, \qquad {}^{n+1}s(i) = \hat{s},$$

where z,s minimize

$$(11)\quad \sum_{j\neq i} q(i,j;z)\big[s_j({}^{n}w(j)-{}^{n}w(i)) + 1 + s_j\log s_j - s_j\big] + d(i;z).$$

Consequently,

$$(12)\quad {}^{n+1}s_j(i) = \exp\{{}^{n}w(i) - {}^{n}w(j)\}, \qquad j\in I.$$

The sequence $\{{}^{n}\lambda,\ n=0,1,\dots\}$ is nonincreasing (see also Remark 1).

3. Iteration procedure

Using (12) we can redefine the algorithm. In the course of the iterations a sequence

$${}^{n}u,\quad {}^{n}\lambda,\quad {}^{n}w(j),\quad j\in I,\quad n=1,2,\dots,$$

is computed. In the step $n \longrightarrow n+1$ one makes

$$(13)\quad {}^{n+1}u(i) = \hat{z}, \quad \text{where} \quad \sum_j q(i,j;\hat{z})\exp\{-{}^{n}w(j)\} = \max.$$

Further, ${}^{n+1}\lambda$ is computed, and ${}^{n}w(j)$, $j\in I$, corrected into

$${}^{n+1}w(j) = {}^{n}w(j) + \Delta w(j)$$

by solving the equations

$$(14)\quad \sum_j q(i,j;{}^{n+1}u(i))\exp\{{}^{n}w(i)-{}^{n}w(j)\}(\Delta w(j)-\Delta w(i)-1) - {}^{n+1}\lambda = 0, \qquad i\in I.$$

The procedure is initialized by a choice ${}^{o}w(j)$, $j \in I$. To determine the auxiliary quantities ${}^{n}w(j)$, $j \in I$, uniquely, one fixes in I a state, say 0, and sets ${}^{n}w(0) = 0$, $n=0,1,\ldots$

To deduce (13) one has to insert from (12) into (11), and to recall (4). Similarly, the substitution from (12) into (10) yields (14).

<u>Convergence.</u> Under the hypothesis

$$(15) \qquad \varepsilon \leqq {}^{n}s_j(i) \leqq \varepsilon^{-1}, \quad n = 1,2,\ldots, \ i,j \in I,$$

for an $\varepsilon > 0$, the convergence of the procedure can be demonstrated as it is done in [1] for semi-Markov processes. But it is to be noted that the Howard algorithm for semi-Markov processes does not coincide in the Markovian case with the above introduced one.

To outline the convergence proof let us consider the controlled process V under the control $({}^{n}u(i), {}^{n}s(i))$. Denote by $C_t = \int_0^t c\,ds$ the total cost accumulated up to time t. Define

$$(16) \qquad M_t = C_t - {}^{n}\lambda\, t + {}^{n}w(V_t) - {}^{n}w(V_o), \quad t \geqq 0.$$

Whenever $V_t = i$, a jump $i \longrightarrow j$ occurs in the infinitesimal interval $(t,t+dt)$ with probability

$$q(i,j;\, {}^{n}u(i))\, {}^{n}s_j(i)dt.$$

Consequently,

$$E\{M_{t+dt} - M_t \mid V_t = i\} = c(i;{}^{n}u(i),\, {}^{n}s(i))dt -$$
$$- {}^{n}\lambda\, dt + \sum_{j \neq i} q(i,j\,;\, {}^{n}u(i))\, {}^{n}s_j(i)({}^{n}w(j) - {}^{n}w(i))dt = 0,$$

in virtue of (10). We conclude that M is a martingale. Similarly, from

$$c(i;{}^{n}u(i),\, {}^{n}s(i)) + \sum_{j \neq i} q(i,j;\, {}^{n}u(i))\, {}^{n}s_j(i)({}^{n-1}w(j) - {}^{n-1}w(i)) -$$
$$- {}^{n-1}\lambda \leqq 0, \quad i \in I,$$

it is derived that

$$(17) \qquad N_t = C_t - {}^{n-1}\lambda\, t + {}^{n-1}w(V_t) - {}^{n-1}w(V_o), \quad t \geqq 0,$$

is a supermartingale.

In state 0 it is ${}^{n}w(0) = {}^{n-1}w(0) = 0$. Introduce the first entrance time into 0

$$\tau = \inf \{ t : V_t = 0 \} .$$

Writing ${}^{n}E_i$ for the mathematical expectation under the condition $V_o = i$, we get from (16), (17)

$$(18) \qquad 0 \leq {}^{n}E_i (M_\tau - N_\tau) = ({}^{n-1}\lambda - {}^{n}\lambda)\, {}^{n}E_i \tau + {}^{n}w(i) - {}^{n-1}w(i).$$

(15) implies that the control parameter space for the process V is compact, and, in virtue of Assumption 1,

$${}^{n}E_i \tau \;, \quad {}^{n}w(i), \qquad n=1,2,\ldots, \qquad i \in I,$$

are bounded sequences. (18) gives

$$(19) \qquad {}^{n}w(i) - {}^{n+k}w(i) \leq ({}^{n}\lambda - {}^{n+k}\lambda) \sup_m {}^{m}E_i \tau$$

for $n,k = 1,2,\ldots$ Since $\{ {}^{n}\lambda, \; n=1,2,\ldots \}$ is a convergent sequence, (19) shows that $\{ {}^{n}w(i), \; n=1,2,\ldots \}$ cannot oscillate as $n \to \infty$. Hence, the limites

$$\lim_{n\to\infty} {}^{n}\lambda \;, \quad \lim_{n\to\infty} {}^{n}w(j), \qquad j \in I,$$

exist and satisfy (9). Moreover, (9) implies

$$\lim_{n\to\infty} {}^{n}\lambda = \hat{\lambda} .$$

<u>Remark 1.</u> Note that

$$0 \leq {}^{n}E_i (M_t - N_t) = ({}^{n-1}\lambda - {}^{n}\lambda) t + O(1), \qquad \text{as} \quad t \to \infty \quad .$$

This yields ${}^{n-1}\lambda \geq {}^{n}\lambda$.

<u>Example 2.</u> The branching Markov process with several types describes the evolution of a population of individuals. The probability that an individual of type i existing at time t undergoes a transformation during the infinitesimal interval $(t, t+dt)$ equals $a_i dt$. The result of this transformation is a random batch of individuals. Its probability distribution depends only on i. Let m_{ij} denote the expected number of individuals j arising from the transformation of an individual i. Further, let $e_i(t)$ be the expected number of individuals i at time t. From the above follows

$$(20) \quad \frac{d}{dt} e_i(t) = \sum_j e_j(t) a_j m_{ji} - e_i(t)a_i, \quad i \in I,$$

where I denotes the set of the types of the individuals.

Set

$$q(i,j) = a_i m_{ij}, \quad i \neq j, \quad q(i,i) = a_i(m_{ii}-1),$$

$$Q = \| q(i,j) \|_{i,j \in I} .$$

From (20) it follows that the principal eigenvalue of Q is the Malthusian parameter of the population's growth (decay). If there are control variables in q, Problem 1 is the problem of maximizing the Malthusian parameter.

4. References

[1] A.Federgruen, H.C.Tijms: The optimality in average cost denumerable state semi-Markov decision problems, recurrency conditions and algorithms. J.Appl.Probability 15 (1978), 356-373.

[2] W.H.Fleming : Logarithmic transformation and stochastic control. Proc.IFIP Working Conference on Recent Advances in Filtering and Optimization, Cocoyoc, Mexico, February 1 to 6, 1982. To appear in Lecture Notes in Control and Information Sciences, Springer-Verlag.

[3] R.A.Howard : Dynamic Programming and Markov Processes, Cambridge (Mass.)-New York,1960.

[4] P.Mandl : An iterative method for maximizing the characteristic root of positive matrices. Rev.roum.Mathématiques pures et appl. XII (1967), 1317-1322.

[5] K.Sladký : Bounds on discrete dynamic programming recursions I, II. Kybernetika (Prague) 16 (1980), 526-547, 17 (1981), 310-328.

ON A CLASS OF LEARNING ALGORITHMS WITH SYMMETRIC BEHAVIOR UNDER SUCCESS AND FAILURE

by

M. R. Meybodi and S. Lakshmivarahan
School of Electrical Engineering and Computer Science
University of Oklahoma, Norman, Oklahoma

1. Introduction:

Learning algorithms have been extensively studied in Mathematical Psychology [13] [14] [18] for decades and more recently in Learning Automata Theory and Mathematical Statistics [1] [2] [8] [11] [16]. In Mathematical Psychology the interest in learning stems from the desire to understand the observed animal learning and associated changes in their behavior. However, in Learning Automata Theory and Mathematical Statistics the aim is to build algorithms that exhibit prespecified behaviour. Our interests, in this paper, are in the latter approach.

We study a new class of absorbing barrier algorithms of the reward-penalty type with identical behaviour under the occurrence of success and failure. Necessary and sufficient conditions for strong absolute expediency of this class of algorithm is obtained.

The concept of absolute expediency was originally introduced by Lakshmivarahan and Thathachar [3] in 1973. Since then this concept has played a major role in the analysis and design of ε-optimal absorbing barrier algorithms. [1] [2] [4] [5] [7]. The results of this paper sheds further light on this concept. It should be interesting to note that ε-optimal non-absorbing learning algorithms have also been recently developed in [15] and discussed extensively in the book [21].

A number of learning algorithms of the reward-penalty type whose behaviour is asymmetrical with respect to the occurrence of success and failure have been extensively studied in the literature - Varshavskii and Vorontsova [10], Fu [8], McMurty and Fu [9], Shapiro and Narendra [20], Viswanathan and Narendra [19], Chandrasekaran and Shen [17], Lakshmivarahan and Thathachar [3] and Sawaragi and Baba [7] to mention a few. For a similar class of general learning algorithms discussed in this paper Aso and Kimura [5] derived necessary and sufficient conditions for absolute expediency [3]. Interestingly Aso and Kimura [5] call this class of algorithms as "Stochastic Vector Automaton" algorithms. Recently for the same class of algorithms considered in this paper Herkenrath, Kalin and Lakshmivarahan [6] derived necessary and sufficient conditions for absorption at the vertices of the unit simplex of proper dimension.

In section 2, the algorithm and the statement of problem are given. Necessary and sufficient conditions for strong absolute expediency and convergence of the algorithm with probability one are established in section 3. ε-optimality is derived in section 4.

2. LEARNING ALGORITHM and STATEMENT OF PROBLEM

There are M ($2 \leq M < \infty$) coins. At time instant k (= 0,1,2....) the coin i ($1 \leq i \leq M$) is chosen for tossing with probability $P_i(k)$ where $P(k) = (P_1(k), P_2(k), \ldots P_M(k))^T, \sum_{i=1}^{M} P_i(k) = 1, 0 \leq P_i(k)$ and T denotes transpose. In tossing, the ith coin falls head (tail) with probability d_i ($c_i = 1 - d_i$). It is assumed that (1) $0 < d_i < 1$, i = 1, 2, ...M, (2) the d_i's are all distinct, that is, $d_i \neq d_j$ for all i and j. (3) d_i's do not depend on k and (4) d_i's are all unknown. The outcome of falling head is called success and falling tail is called failure. Let $D = (d_1, d_2 \ldots d_M)^T$ and without loss of generality assume

$$d_1 > d_2 > d_3 \ldots > d_M \tag{1}$$

Any vector D that satisfies the above conditions is called an admissible D. The average probability of success at stage k denoted by $\eta(k)$ is

$$\eta(k) = \sum_{i=1}^{M} P_i(k)\, d_i \tag{2}$$

Our basic problem is to make the average probability of success as close to its maximum (that is, d_1) as possible. The primary interest in and the challenge of this problem arise basically from the fact that the success probabilities (d_i's) of the coin are unknown.

As a first step towards the solution to this problem, in this paper, we propose to change p(k) in a learning algorithm. The key idea of the learning algorithm may be stated in words as follows: Increase (decrease) the probability of choosing the ith coin at time (k + 1), if it was chosen for tossing at time k and the toss resulted in success (failure). In particular, let

$$S_M = \{ P \mid P = (P_1, P_2, \ldots P_M)^T, 0 \leq P_i, \sum_{i=1}^{M} P_i = 1\}$$

be the M-dimensional unit simplex and let

$$V_M = \{ e_i \mid i = 1, 2, \ldots M\} \text{ where } e_i = (0, 0, \ldots 1, \ldots 0)^T$$

be the ith unit i vector of dimension M. Further, let $S_M^o = \{ P \mid P = (P_1\ P_2, \ldots P_m)^T\ 0 < P_i, \sum_{i=1}^{M} P_i = 1\}$ Clearly, V_M corresponds to the corners of vertices of S_M.
Let $f_s^i [\cdot], g_s^i [\cdot]$ be continuous functions such that

$$f_s^i : S_M \to [0,1],\ g_s^i : S_M \to [0,1],\ i, s = 1, 2, \ldots M$$

Formally the above algorithm may be stated as follows

$$\left.\begin{aligned} P_s(k+1) &= P_s(k) - \theta f_s^i [P(k)], s \neq i \\ P_i(k+1) &= P_i(k) + \theta f_i^i [P(k)] \end{aligned}\right\} \text{ if the toss of coin i resulted in success}$$

and

$$\left.\begin{aligned} P_i(k+1) &= P_i(k) - \theta g_i^i [P(k)] \\ P_s(k+1) &= P_s(k) + \theta g_s^i [P(k)], s \neq i \end{aligned}\right\} \text{ if the toss of coin i resulted in failure} \tag{3}$$

where $0 < \theta \leq 1$ is called the <u>step length parameter</u> and the following consistency condition (4):

$$\text{either } f_s^i\,[.] \equiv 0 \text{ for all } p \in S_M$$
$$\text{or } f_s^i\,[p] \leq P_s \quad s \neq i \text{ and } f_i^i\,[p] = \sum_{s \neq i} f_s^i\,[p] \tag{4}$$

and

$$\text{either } g_s^i\,[p] \equiv 0 \text{ for all } p \in S_M$$
$$\text{or } g_s^i\,[p] \leq P_s \text{ and } g_i^i\,[p] = \sum_{s \neq i} g_s^i\,[p]$$

for all i, s = 1, 2, ...M,

imply that $p(k + 1) \in S_M$ if p(k) does. However in order to make the algorithm non trivial and interesting either $f_s^i \equiv 0$ or $g_s^i \equiv 0$ but not both.

<u>Remark 1.</u> If the coin i is chosen for tossing, it is clear from (3) that success (failure) increases (decreases) the probability of its choice. The increase and decrease are called reward and penalty and hence (3) is called reward-penalty algorithm. If $f_s^i \not\equiv 0$ but $g_s^i \equiv 0$ then (3) is called reward-inaction algorithm, if $f_s^i \equiv 0$ but $g_s^i \not\equiv 0$ then it is called inaction- penalty algorithm

Since the vector D, of success probabilities and the functions $f_s^i\,[.]$ and $g_s^i\,[.]$ (for all i and s) are independent of k, $\{p(k)\}, k \geq 0$ is clearly a discrete time Markov process with stationary transition function over the state space S_M. To quantify the behaviour of the process {p(k)} , and hence of $\{\eta(k)\}$ we introduce the following definitions. Let I = {1,2,....M } E = {success, Failure}. The algorithm (3) defines a mapping T: $S_M \times I \times E \to S_M$ where

$$P(k+1) = T\,[p(k), i(k), e(k)] \tag{5}$$

$i(k) \in I$ is the coin chosen for tossing and $e(k) \in E$ is the outcome of the toss of that coin at time k.

<u>Definition 1:</u> A state $p \in S_M$ is said to be <u>absorbing state</u> if and only if p = T [p,i,e,] with probability one (6)

<u>Definition 2:</u> A learning algorithm T is said to be <u>absorbing</u> if and only if there is at least one absorbing state.

For reasons that will become apparent in this paper our interests are in the class of learning algorithms for which V_M is the only set of absorbing states. We call such an algorithm <u>absorbing barrier</u> algorithms

<u>Definition 3:</u> A learning algorithm T is said to be

a) <u>optimal</u> if

$$\lim_{k \to \infty} E\,[\eta(k)] = d_1 \tag{7}$$

b) <u>ε-optimal</u> if for all $p(0) = p \in S_M^o$

$$\lim_{k \to \infty} |\,E\,[\eta(k)] - d_1| < \varepsilon\,,\ \varepsilon > 0 \tag{8}$$

c) <u>absolutely expedient</u> if

$$E\,[\eta(k+1) \;|\, p(k) = p] \geq \eta(k) \text{ with probability one} \tag{9}$$

for all admissible D (satisfying the assumptions given at the beginning of this section) and for all $p \in S_M$, with strict inequality in (9) holding good for all $p \in S_M^o$.

d) strongly absolutely expedient if

$$E[\eta(k+1) \mid p(k) = p] \geq \eta(k) \text{ with probability one} \quad (10)$$

for all admissible D and for all $p \in S_M$ with strict inquality in (10) holding good for all $p \in (S_M - V_M)$.

STATEMENT OF PROBLEM: Our aim in this paper is to find conditions for ε-optimality of the general class of learning algorithm given in (3).

Remark 2: Algorithm (3) is a generalization of many known algorithms in the sense that the functions in (3) used for updating the probabilities depend on the coin chosen as well. Such generalized algorithms are considered in [5] and [6]. In most of the earlier papers the functions that are used in updating are independent of the coin that is chosen for tossing, [1] [2] [3] [4] [7] [8] [9] [10] [12] [13].

3. Convergence with probability one:

As a first step towards ε-ptimality, in this section we derive conditions on the algorithm such that the Markov process $\{p(k)\}$ $k \geq 0$ converges with probability one. To this end we begin be rewriting the consistency conditions (4) in a form more suitable for our analysis. Let

(C.1) $$f_i^i[p] = \alpha[i,p](1 - p_i), \quad f_s^i[p] = \beta[i,s,p]\, p_s$$

and

$$\sum_{s \neq i} \beta[i,s,p]\, p_s = \alpha[i,p](1 - p_i)$$

and

(C.2) $$g_i^i[p] = \gamma[i,p]\, p_i, \quad g_s^i[p] = \delta[i,s,p](1 - p_s)$$

and

$$\gamma[i,p]\, p_i = \sum_{s \neq i} \delta[i,s,p](1 - p_s)$$

where $\alpha, \gamma : I \times S_M \to [0,1]$ and $\beta, \delta : I \times I \times S_M \to [0,1]$.

The basic rule that governs the choice of functions in the above form is that if a term is substracted from p_j then it is made proportional to p_j and if a term is added to p_j then it is made proportional to $(1 - p_j)$ irrespective of which coin is chosen for the toss and whether the toss results in success or failure.

Remark 3: Our choice of the functions $g_s^i[.]$ is quite untraditional in the sense that in all most all the papers in Mathematical Psychology [18] [17] [13] [14] $g_s^i[p]$ is made proportional to p_s for all i and s, $s \neq i$. Also in almost all the papers on Learning Automata [3] [4] [5] [7] $g_i^i[p]$ is made proportional to $(1 - p_i)$ and $g_s^i[p]$ is made proportional to p_s for all $s \neq i$. Because of this there is a disparity in the behaviour of the algorithm (3) under success and failure. However, our present choice of functions $g_s^i[.]$ for all i and s given in (C.2) induce identical behaviour of the algorithm (3) under success and failure.

The following theorem is immediate

Theorem 1: The learning algorithm (3) is an absorbing barrier learning algorithm if and only if (A.1) and (A.2) hold true.

(A.1) For all $P \varepsilon S_M - V_M$ there exists $1 \leq s \leq M$ such that $[\alpha [s,p] + \gamma[s, p]] p_s > 0$

(A.2) For all $1 \leq s \leq M$, $\gamma [s, e_s] = o$

Proof: Refer [6]

A typical choice of functions in the algorithm (3) that satisfies the conditions (C.1) - (C.2) and (A.1) - (A.2) are given in the following example.

$$\text{Let } 0 < C_1, C_2 < 1.$$

$$\beta [i,s,p] = C_1 \ (1 - p_i) (1 - p_s), \ \alpha[i, p] = C_1 \sum_{s \neq i} p_s (1 - p_s)$$

$$\delta [i, s, p] = C_2 \ p_i \ p_s^2 (1-p_i), \ \gamma[i, p] = C_2 (1 - p_i) \sum_{s \neq i} p_s^2 (1-p_s)$$

for all i, s = 1, 2, ...M.

Remark 4: The demonstration of the symmetry in the properties of the algofithm (3) under success and failure alluded to inremark 3 is evidenced by theorem 1. Notice that (3) is an absorbing barrier learning algorithm if $\alpha[i,p] \not\equiv 0$ and $\gamma[i,p] \equiv 0$ or $\alpha [i,p] \equiv 0$ and $\gamma [i,p] \not\equiv 0$ or both $\alpha[i,p]$ and $\gamma[i,p] \not\equiv 0$. This is in sharp contrast with the properties of the currently available absolutely expedient learning algorithm [3] [4] [5] [7] wherein the reward-penalty and reward-inaction algorithms are absorbing barrier type but the inaction-penalty is not. In fact in all the inaction-penalty algorithms of the absolutely expedient type known so far [3], every state in δS_M is an absorbing state. Our modified definition of strong absolute expediency is in fact motivated by the existence of the absorbing barrier algorithms of the reward-penalty, reward-inaction and inaction-penalty types.

Remark 5: For some $1 \leq j \leq M$ if $p_j(k) = 0$, then it follows from (3) and (C.1) - (C.2) that $p_j(k^*) = 0$ for all $k^* \geq k$. In other words, during learning process if p(k) reaches the boundary $p_j = 0$ of the simplex S_M, then p(k) will continue to reamin in that boundary.

Henceforth in this paper we will only be concerned with the absorbing barrier learning algorithms, that is, algorithm (3) under the conditions (A.1) - (A.2) of theorem 1.

Theorem 2: Necessary and sufficient conditions for the absorbing barrier learning algorithm (3) to be strongly absolutely expedient are:

(S.1)
$$\sum_{j\neq i} p_j \ \beta [i, j, p] = \sum_{j\neq i} p_j \ \beta [j, i, p]$$

and

(S.2)
$$\sum_{j\neq i} p_i (1 - p_j) \ \delta [i, j, p] = \sum_{j\neq i} p_j (1-p_i) \ \delta [j, i, p]$$

for all i = 1, 2, ...M

Sufficiency: Define $\delta x(k) = x(k + 1) - x(k)$ and let

$$\Delta\eta (k) = E [\delta\eta (k) \mid p (k)] = \sum_{i=1}^{M} E[\delta P_i (k) \mid p(k)] d_i \quad (11)$$

It can be seen by direct computation that

$$E[\delta p_i(k) \mid p(k) = p] = p_i(1-p_i)\, d_i\, \alpha(i,p) - p_i^2\, c_i\, \gamma[i,p] - \sum_{j\neq i} p_j\, p_i\, d_j\, \beta[j,i,p] + \sum_{j\neq i} p_j(1-p_i)\, c_j\, \delta[j,i,p] \tag{12}$$

Substituting (12) in (11) and inview of (C.1) and (C.2) we obtain

$$\Delta\eta(k) = \Delta\eta_1(k) + \Delta\eta_2(k) \tag{13}$$

where

$$\Delta\eta_1(k) = \sum_{i=1}^{M} p_i d_i^2 \sum_{j\neq i} p_j\, \beta[i,j,p] - \sum_{i=1}^{M} p_i d_i \sum_{j\neq i} p_j d_j\, \beta[j,i,p] \tag{14}$$

and

$$\Delta\eta_2(k) = -\sum_{i=1}^{M} p_i d_i c_i \sum_{j\neq i} \delta[i,j,p]\,(1-p_j) + \sum_{i\neq 1}^{M} (1-p_i) d_i \sum_{j\neq i} p_j c_j\, \delta[j,i,p] \tag{15}$$

After simplification it can be shown that [23]

$$\Delta\eta_1(k) = 1/2 \sum_{i=1}^{M} \sum_{j>i} p_i p_j (d_i - d_j)^2 \{\beta[i,j,p] + [j,i,p]\}$$

$$\geq o \text{ with equality holding only if } p \in V_m \tag{16}$$

Similarly it can be shown that [23]

$$\Delta\eta_2(k) = 1/2 \sum_{i=1}^{M} \sum_{j>i} p_i (1-P_j)(d_i - d_j)^2 \{\delta[i,j,p] + \delta[j,i,p]\} \tag{17}$$

$$\geq o \text{ with equality holding good only if } p\ V_m$$

From (16) and (17) sufficiency follows.

<u>Necessity</u>: $\Delta\eta(k)$ can be represented as a quadratic and linear term in the vector D as follows:

$$\Delta\eta(k) = D^T A D + D^T B \tag{18}$$

where $A = [A_{ij}]$ and $B = [B_1, B_2, \ldots. B_m]^T$ with

$$A_{ii} = P_i(1-P_i)\,\alpha[i,P] + P_i^2\,\gamma[i,p]$$

$$A_{ij} = -[P_i P_j\,\beta[j,i,p] + (1-P_i)\,P_j\,\delta[j,i,p]]$$

and

$$B_i = P_i^2\,\delta[i,p] + (1-P_i)\sum_{j\neq i} P_j\,\delta[j,i,p]$$

for all $i,j = 1,2,\ldots M$

From the definition of strong absolute expediency it follows that $\Delta\eta(k)$ attains its minimum value zero either when all d_i are equal or when $p \in V_m$. Since every member of V_m is absorbing, on V_m, it is easily seen that, $\Delta\eta(k)$ attains its minimum value for all admissible D vectors. In the following we shall derive conditions for the minimum of $\Delta\eta(k)$ when $d_i = d$ for all $i = 1,2,\ldots\ldots,M$, $0 < d < 1$. Necessary conditions for minimum are obtained by setting the derivative of $\Delta\eta(k)$ (with respect to d_i) at the point $d_i = d$ for all $i = 1,2,\ldots\ldots,M$ equal to zero, that is,

$$\left.\frac{\partial\Delta\eta}{\partial d_i}\right|_{d_i=d} = 0 \text{ for all } i = 1,2,\ \ldots.M \tag{19}$$

From (18), the equation (19) take the form

$$(A + A^T)\, d\, \underline{1} + B = o \tag{20}$$

where $\underline{1} = (1,1,\ldots.1)^T$ is an M dimensional column vector of all ones.

Rewriting (20) we get

$$d\,K_i + L_i = 0 \tag{21}$$

for all i = 1,2,....,M and all $o < d < 1$

where

$$K_i = P_i\,(1 - P_i)\ \alpha[i,p] + P_i^2\ \gamma[i,p]$$
$$- P_i \sum_{j\neq i} P_j\ \beta[j,i,p] - (1-p) \sum_{j\neq i} P_i\ \delta[j,i,p]$$

and

$$L_i = P_i^2\ \gamma[i,p] + (1-p_i) \sum_{j\neq i} P_j\ \delta[j,i,p].$$

(21) is true for all $0 < d < 1$ only if $L_i = 0$ and $K_i = 0$ for all i = 1,2,..., M.

$L_i = 0$ leads to the condition

$$P_i^2\ \gamma[i,p] = (1 - P_i) \sum_{j\neq i} P_j\ \delta[j,i,p]. \tag{22}$$

Using (C.2), from (22) we obtain

$$P_i \sum_{j\neq i} (1-P_j)\ \delta[i,j,p] = (1 - P_i) \sum_{j\neq i} P_j\ \delta[j,i,p]. \tag{23}$$

which in fact is (S.2). Substituting (23) in $K_i = 0$, we obtain

$$P_i\,(1-P_i)\ \alpha[i,p] = P_i \sum_{j\neq i} P_j\ \beta[j,i,p].$$

Once again, using (C.1) we get $P_i \sum_{j\neq i} P_j\ \beta[i,j,p] = P_i \sum_{j\neq i} P_j\ \beta[j,i,p]$

which is the same as (S.1). Hence the theorem.

Corollary 1: The Markov process $\{p(k)\}\ k \geq 0$ as generated by the algorithm (3) under the conditions (C.1) - (C.2), (A.1) - (A.2) and (S.1) - (S.2) converges to V_M with probability one.

Proof: Since $\{p\ (k)\}$ is a Markov process, from theorem (2) we get

$$E\,[\delta\eta\ (k) \mid P\ (r):\ o \leq r \leq k] = E\,[\delta\eta\ (k) \mid P\ (k)] \geq 0$$

This inturn implies that $\{\eta(k)\}\ k \geq 0$ is a submartingale [22]. By martingale theorem $\lim_{k\to\infty} \eta(k)$ exists and hence $\lim_{k\to\infty} p(k) = p^*$ exists with probability one. As $\delta\eta(k) = 0$ only on V_M, it follows that $P^* \in V_M$ with probability one.

4. ε - Optimality: In the previous section it was established that $P^* \in V_M$ with probability one. In this section we set out to quantify the distribution of P*.

To this end, define

$$\Gamma_i\,(p) = \text{Prob}\,[P^* = e_i \mid P\ (0) = P] \tag{24}$$

for i = 1,2,..., M. Notice $\sum_{i=1}^{M} \Gamma_i\,(p) = 1$ for all $p \in S_M$

In view of the corollary 1 and (24) we obtain, for all $p(0) = p \in S_M^o$

$$\lim_{k\to\infty} E\,[\ \eta(k)] = \sum_{i=1}^{M} \Gamma_i(p)\ d_i \tag{25}$$

To compute $\Gamma_i\,(p)$ we need the following: Let C $[S_M]$ be the class of all continuous functions from S_M to the real line. If $f\,(.) \in C[S_M]$ define the operator U as follows:

$$U\,f\,(p) = E\,[f\,(p\,(k + 1)) \mid p\,(k) = p]$$

Clearly the operator U is linear and positive [12]

Definition 4: A function $f: \to S_M$ real line is called super regular (regular, sub regular) if

$$f(p) \geq (=, \leq) \; U f(p)$$

for all $p \in S_M$.

With these preliminaries, we now state two important propositions that lead an algorithm for quantifying $\Gamma_i(p)$.

Proposition 1: $\Gamma_i(p)$ is the only continuous solution of the functional equation.

$$U \Gamma_i(p) = \Gamma_i(p) \tag{26}$$

satisfying the boundary condition

$$\Gamma_i(e_i) = 1 \text{ and } \Gamma_i(e_j) = 0 \text{ for all } i,j,i \neq j$$

Notice $\Gamma_i(p)$ satisfying (26) by definition is a regular function. This functional equation is extremely difficult to solve. Hence in the following we establish upper and lower bounds on $\Gamma_i(p)$.

Proposition 2: Let $f_i(p) \in C[S_M]$ be super (sub) regular function with $f_i(e_i) = 1$ and $f_i(e_j) = 0$ for all i,j, $i \neq j$ then

$$f_i(p) \geq (\leq) \quad \Gamma_i(p)$$

The proof of these propositions are rather involved and we refer the reader to Norman [12] for an elegant proof.

Thus, if we can find two functions $h_i^{(1)}(p)$ and $h_i^{(2)}(p)$ which are super and sub-regular functions respectively and satisfying the boundary conditions

$$h_i^{(1)}(e_i) = h_i^{(2)}(e_i) = 1, \; h_i^{(\ell)}(e_j) = 0 \text{ for all } j \neq i, \; \ell = 1,2.$$

then from proposition 2 it follows that $h_i^{(2)}(p) \leq \Gamma_i(p) \geq h_i^{(1)}(p)$

Consider a function

$$\Psi_i[x_i, p] = \text{ExP}[-x_i p_i/\theta]$$

where $x_i > 0$ is a parameter and $\text{ExP}[\chi] = e^{\chi}$

Recognizing the fact that

$$\emptyset[X_i, p] = \frac{1 - \text{ExP}[-x_i p_i/\theta]}{1 - \text{ExP}[-x_i/\theta]}$$

is sub (super) regular whenever $\Psi_i[X_i, P]$ is super (sub) regular, it follows from propositions 2 that

$$\emptyset[Y_i, P] \leq \Gamma_i(p) \leq \emptyset[Z_i, p]$$

where Y_i, Z_i are two constants such that $\emptyset[Y_i, p]$ and $\emptyset_i[Z_i, p]$ are sub and super regular function respectively. Note that function $\emptyset_i[x_i, p]$ depends on just one compenent of P and thus leads to conservative results. However, it gives rise to expressions which are easily manageable.

The problem of getting bounds on $\Gamma_i(p)$ now reduces to one of finding two positive constants Y_i and Z_i such that $\emptyset_i[Y_i, p]$ is sub and regular and $\emptyset_i(Z_i, p)$ is super regular. Further, from (25) and the inequality (1) it is clear that for ε - optimality we need to concentrate only (on the lower bound) on $\Gamma_i(p)$.

It can be seen that (by dropping the subscript 1 from x, for convenience)

$$U \Psi_1[x, p_1] - \Psi_1[x, p_1] = -x F_1[x, p] \Psi_1[x, p]$$

where

$$F_1[x, p] = p_1 d_1 \alpha[1, p](1 - p_1) V[-x \alpha[1, p](1 - p_1)]$$

$$-p_1 C_1 \, V[1,p] \, p_1 \, V[x\,\gamma\,[1,p]\,p_1]$$

$$-\sum_{j\neq 1} p_j \, d_j \quad \beta[j,1,p] \, p_1 \quad V[x\,\beta[j,1,p]\,p_1]$$

$$+\sum_{j\neq 1} p_j C_j \quad \delta[j,1,p] \, (1-p_1) \quad V[-x\,\delta[j,1,p]\,(1-p_1)]$$

and V [z] = (ExP (z) - 1)/z if z ≠ 0 and 1 if z = 0.

Clearly, F_1 [x,p] ≥ 0 implies $\emptyset_1$ [x,p] regular.

Since $\alpha[i,p]$, $\gamma[i,p]$, $\delta[i,j,p]$ and $\beta[i,j,p]$ are all bounded above by unity and as the V [.] is strictly monotonically increasing, it follows

F_i [x,p] $\geq$ 0 if

$$G[x,p] \triangleq \frac{V[-x\,(1-p_1)]}{V[x\,p_1]} \geq \frac{p_1^2 \, C_1 \, \gamma[1,p] + \sum_{j=1} p_1 p_j d_j \, \beta\,[j,i,p]}{\sum_{j\neq 1} p_j \, (1-p_1) \, C_j \, \delta[j,1,p] + p_1 \, (1-p_1) \, d_1 \, \alpha[1,p]}$$

It can be shown using the properties of the V [.] function [4] [12] that

$$G\,[x,p] \geq \frac{1}{V\,[x]}$$

Further, using (C.1) - (C.2) and (S.1) - (S.2) it can be seen that

$$e^* \triangleq \frac{C_1 \, A + d_2 B}{C_M \, A + d_1 B} \geq \frac{p_1^2 \, C_1 \quad \gamma[1,p] + \sum_{j\neq 1} \quad p_1 p_j d_j \, \beta[j,1,p]}{\sum_{j\neq 1} p_j \, (1-p_1) \, C_j \quad \delta[j,1,p] + p_1 \quad (1-p_1) \, d_1 \quad \alpha(1,p]}$$

where $A = \sum_{j\neq 1} p_j \, (1-p_1) \, C_j \quad \delta\,[j,1,p]$ and $B = \sum_{j\neq 1} p_1 p_j d_j \quad \beta[j,1,p]$

In view of the inequality (1) it follows that $e^* \leq 1$. From (27) and (28)

$$F_1 \, [x,p] \geq 0 \text{ if } \frac{1}{V\,[x]} \geq e^* \qquad (29)$$

Since $e^* < 1$; $V\,[x] = \frac{1}{e^*}$ has an unique solution x = y > 0 so that $\phi[y,p]$ is subregular.

Remark 7: For the reward-incation algorithm (refer remark1) e* reduces to $\frac{d_2}{d_1}$ and for the inaction - penalty algorithm e* reduces to $\frac{C_1}{C_M}$ Thus in either of these special cases there exists an unique solution x=y > 0 for the equation $V\,[x] = \frac{1}{e^*}$ and hence in both these special cases there exists a lower bound on Γ_1 (p). This is in sharp contrast with the existing results in the literature where in for the inaction-penalty (absolutely expedient) algorithms [4] no lower bound on Γ_1 (p) has been established. One of the reasons for this anamoly is that none of the currently available inaction-penalty (absolutely expedient) algorithms [4] are of the absorbing barrier type.

Having established the lower bound on Γ_1 (p) we now state our main result.

Theorem 3: For every $\varepsilon > o$ and $p(0) = p \in S_M^o$ there esists $0 < \theta^* < 1$ such that for $0 < \theta < \theta^*$, the learning algorithm under the conditions (C.1) - (C.2), (A.1) - (A.2) and (S.1) - (S.2) is such that

$$\lim_{k\to\infty} \left| E[\eta\,(k)] - d_1 \right| < \varepsilon \qquad (30)$$

Proof: From corollary 1 it follows that the limit in the left hand side of (30) exists. From (25)

$$\lim_{k\to\infty} E[\eta(k)] = \Gamma_1(p) d_1 + \sum_{j=1} \Gamma_j(p)\, d_j$$

That is,

$$\lim_{k\to\infty} E[\eta(k) - d_1] \leq (1-\Gamma_1(p)) \quad (d_2 - d_1)$$

Since $\emptyset[7,p] \leq \Gamma_1(p)$, Combining this with the above,

$$\lim_{k\to\infty} |E[\eta(k)] - d_1| \leq [1-\emptyset_1[y,p]] \quad |(d_2 - d_1)|$$

For any $p \in S_M^o$, we know that

$$\lim_{\theta\to 0} \emptyset_1[x,p] = 1$$

Combining these we see that for any $\varepsilon > 0$ there exists $0 < \theta^* < 1$ such that for all $0 < \theta < \theta^*$

$$\lim_{k\to\infty} |E[\eta(k)] - d_1| < \delta \quad |(d_2 - d_1)|$$

The theorem follows by choosing $\delta = \varepsilon / |(d_2 - d_1)|$.

5. Conclusions: A new class of absorbing barrier absolutely expedient learning algorithms, whose behaviour under the action of success and failure are identical, is introduced. Conditions for the ε - optimality of this class of learning algorithms are derived.

6. References:

[1] K.S. Narendra and M.A.L. Thatachar. "Learning Automata - A survey." IEEE Transactions on Systems, Man, and Cybernetics. Vol. 4, pp. 323-334, 1974.

[2] K.S. Narendra and S. Lakshmivarahan. "Learning Automata - A critique." Journal of Cybernetics and Information Sciences - Special Issue on Learning Automata, Vol. 1, pp. 53-66, 1978.

[3] S. Laksmivarahan and M.A.L. Thatachar. "Absolutely Expedient Learning Algorithms for Stochastic Automata." IEEE Transactions on Systems, Man and Cybernetics, Vol. 3, pp. 281-286, 1973.

[4] S. Laksmivarahan and M.A.L. Thathachar. "Bounds on the Probability of Convergence of Learning Automata.: IEEE Transactions on Systems, Man and Cybernetics. Vol. 6, pp. 756-763, 1976.

[5] H. Aso and M. Kimura. "Absolute Expediency of Learning Automata." Information Sciences, Vol. 17, pp. 91-112, 1979.

[6] U. Herkenrath, D. Kalin and S. Lakshmivarahan. "On a General Class of Absorbing Barrier Learning Algorithms." School of EECS Technical report - 8001 University of Oklahoma, March, 1980. (Also in Information Sciences Vol. 24 pp. 255-263, 1981.

[7] Y. Sawaragi and N. Baba. "Two ε- optimal Non-Linear Reinforcement Schemes for Stochastic Automata." IEEE Transactions on Systems, Man and Cybernetics. Vol. 4, pp. 126-131, 1974.

[8] K.S. Fu. "Stochastic Automata Models for Learning Systems." in Computers and Information Sciences II (Ed) by J.T. Tou, Academic Press, 1967.

[9] G.J. McMurtry and K.S. Fu. "A Variable Structure Automaton Used in Multimodal Search Technique. " IEEE Transactions on Automatic Control, Vol 11, pp. 379-387, 1966.

[10] V.I. Varshavskii and I.P. Vorontsova. "On the Behaviour of Stochastic Automata with Variable Structure, " Automation and Remote Control, Vol. 24, pp. 327-333, 1963.

[11] M.L. Tsetlin. "Automaton Theory and Modelling of Biological Systems" Academic Press, 1973.

[12] M.F. Norman. "On a Linear Model with Two absorbing Barriers." Journal of Mathematical Psychology. Vol. 5, pp. 225-241, 1968.

[13] M.F. Norman. Markov Processes and Learning Models. Academic press, New York, 1972.

[14] M. Iosifescu and R. Theodorescu. "Random Process and Learning." Springer Verlag, 1969.

[15] S. Lakshmivarahan. " ε - optimal Learning Algorithms - Non Absorbing Barrier Type." Technical Report, EECS 7901, School of EECS, University of Oklahoma, February, 1979.

[16] I.H. Witten. " The Apparent Conflict Between Estimation and Control - A Survey of Two Aremd-Bandit Problem." Journal of Franklin Institute Vol. 301, pp. 161-189, 1976.

[17] B. Chandrasekaran and D.W.C. Shen. "On Expediency and Convergence in Variable Sturcture Automata." IEEE Transactions on Systems Science and Cybernetics. Vol. 4, pp. 52-60, 1968.

[18] R.R. Bush and F. Mosteller. Stochastic Models for Learning. John Wiley, New York, 1958.

[19] R. Viswanathan and K.S. Narendra. "Expedient and Optimal Variable Structure Stochastic Automata." Dunharm Lab. TR-37, 1970. Yale University.

[20] I.J. Shapiro and K.S. Narendra. "Use of Stochastic Automata for Parameter Self Optimisation with Multimodal Performance Criteria." IEEE Transactions on Systems, Man and Cybernetics. Vol. 5, pp. 352-360, 1969.

[21] S. Lakshmivarahan. Learning Algorithms: Theory and Applications. Springer Verlag, New York, 1981.

[22] J.L. Dobb. Stochastic Processes. John Wiley, 1955.

[23] M.R. Meybodi and S. Lakshmivarahan. " ε- Optimality of a general class of absorbing barrier learning algorithms." School of EECS Technical Report, August 1981. University of Oklahoma, Norman, Oklahoma, U.S.A.

CONVERGENCE OF A GENERAL STOCHASTIC APPROXIMATION PROCESS UNDER CONVEX CONSTRAINTS AND SOME APPLICATIONS

J.M. Monnez

Université de Nancy II - I.U.T. Informatique
2 bis, bd Charlemagne - 54000 Nancy - France

I. INTRODUCTION

Albert and Gardner applied in [1] the stochastic approximation methods to the estimation of the vector parameter θ ($\theta \in \mathbb{R}^k$) of a regression model

$$v_n = g_n(\theta) + r_n$$

where for $n \geqslant 1$, v_n is an observable random variable in $\mathbb{R}$, g_n a known function from $\mathbb{R}^k$ into $\mathbb{R}$, r_n a random variable in $\mathbb{R}$ whose expectation is 0.

Albert and Gardner used the stochastic approximation process $(X_n,\ n \geqslant 1)$ in $\mathbb{R}^k$ to estimate θ :

$$X_{n+1} = X_n - \alpha_n(X_1, X_2, -, X_n)(g_n(X_n) - v_n)$$

where for $n \geqslant 1$, α_n is a measurable function from $\mathbb{R}^{kn}$ into $\mathbb{R}^k$.

In the case where θ is a point of a non-empty closed convex subset K of $\mathbb{R}^k$ (for instance in the cases of regression with positive real parameters or of ridge regression) it seems reasonable to consider the process

$$X_{n+1} = P(X_n - \alpha_n(X_1, X_2, -, X_n)(g_n(X_n) - v_n))$$

where P is the projection operator on K. But Albert and Gardner did not succeed in proving its convergence and proposed to modify the estimation scheme in the following manner : the real data v_n are grouped together in batches ; the size of the nth batch is denoted m_n ; thus the following regression model is now considered :

$$V_n = G_n(\theta) + R_n$$

where for $n \geqslant 1$, V_n is an observable random variable in $\mathbb{R}^{m_n}$, G_n a known function from $\mathbb{R}^k$ into $\mathbb{R}^{m_n}$ and R_n a random variable in $\mathbb{R}^{m_n}$ whose expectation is 0 ; the following process is used :

$$X_{n+1} = P(X_n - a_n(X_1, X_2, -, X_n)(G_n(X_n) - V_n))$$

where for $n \geqslant 1$, a_n is a measurable function from K^n into $\mathbb{R}^{k \times m_n}$ (the set of the $k \times m_n$ real matrices). Albert and Gardner proved that X_n converges to θ in mean square when $m_n \geqslant k$ for all n.

The results given in the present paper permit us to get theorems of convergence with probability one of X_n to θ ; the condition $m_n \geqslant k$ is not always required : for instance, we show that we may take $m_n = 1$ in the linear regression case ; moreover the observations V_n may not be uncorrelated.

These theorems are corollaries of theorems of convergence with probability one of a more general stochastic approximation process which are given here. The present work includes an extension of a theorem of Ruppert concerning a dynamic stochastic approximation process [8] ; it too extends to a more general case the study of the convergence of a stochastic approximation process with correlated observations defined by Ljung [5].

The process is defined in section 2. The convergence with probability one is studied in section 3. We use to this end the convergence lemma of the non-negative almost supermartingales of Robbins and Siegmund [7]. We show off the structure of the results obtained by this method : first the lemma is used to prove that the process converges ; then the convergence to the correct point is established under different sets of assumptions. Thus we get three theorems of convergence with probability one. We give the application to the estimation of the vector parameter of a linear regression model in section 4.

The detailed proofs can be found in [6].

2. DEFINITION OF THE STOCHASTIC APPROXIMATION PROCESS

2.1. The general process

The σ-field used on every Hilbert space in this study is generated by the open sets. Let H be a separable real Hilbert space. Inner product and norm in H are denoted respectively $\langle . , . \rangle$ and $\| . \|$. Let K be a non-empty closed convex subset of H and P the projection operator on K.
For $n \geqslant 1$, let H_n' be a separable real Hilbert space.
Let $(\Omega, \mathcal{A}, \mu)$ be a probability space. For $n \geqslant 1$, let Y_n, Z_n and β_n be H_n' valued random variables defined on $(\Omega, \mathcal{A}, \mu)$.
For $n \geqslant 1$, let A_n be a measurable function from $K^n \times H_n'$ into H ; $A_n (x_1, x_2, - , x_n ; y)$ is supposed to be linear with respect to y $(x_1, x_2, - , x_n \in K ; y \in H_n')$.

We consider the process $(X_n, n \geqslant 1)$ in K :

$$X_{n+1} = P (X_n - A_n (X_1, X_2, - , X_n ; Y_n + Z_n + \beta_n))$$

We suppose that $E [\| X_1 \|^2] < \infty$.
Let F_n be the sub-σ-field of $\mathcal{A}$ generated by the events before the time n.

2.2. Some special cases

We give here some special cases in order to illustrate the meaning of the process.

2.2.1. First special case

Choose $K = H = \mathbb{R}^k$, $\forall n$, $H_n' = \mathbb{R}^k$; $\forall n$, $Z_n = 0$, $\beta_n = 0$; $\forall n$, $A_n (X_1, X_2, -, X_n ; Y_n) = a_n Y_n$ ($a_n \in \mathbb{R}^+$).

So we have the process : $X_{n+1} = X_n - a_n Y_n$.

Suppose that for all n $E [Y_n \mid F_n] = M_n (X_n)$ a.s., where M_n is an unknown measurable function from $\mathbb{R}^k$ into $\mathbb{R}^k$. Y_n is a noised observation at time n of $M_n (X_n)$.
Suppose that the zero θ_n of M_n exists and is unique and that $\theta = \lim_{n\to\infty} \theta_n$ exists.
The aim is to get an estimation of θ ; the convergence of the process (X_n) to θ can be proved under some conditions.

This type of process was studied by Burkholder [3]. Special cases of it are the well-known processes of Robbins-Monro and Kiefer-Wolfowitz.

2.2.2. Second special case

Consider the preceding process with for all n $M_n = M$. Now suppose that $Y_n = M(X_n) + Z_n + \beta_n$ where $E [Z_n + \beta_n \mid F_n] \neq 0$. Z_n and β_n are additive noises on $M(X_n)$; the assumptions concerning Z_n will be different from those concerning β_n.
This model can be used in cases where the observations Y_n are correlated.
So we have the process :

$$X_{n+1} = X_n - a_n (M(X_n) + Z_n + \beta_n)$$

which is a special case of the process 2.1. The study of its convergence was made by Ljung in the case where the zero of M may not be unique [5].

2.2.3. A dynamic stochastic approximation case

Choose $\forall n$, $H_n' = \mathbb{R}$; $\forall n$, $Z_n = 0$, $\beta_n = 0$. For $n \geqslant 1$, let M_n be a measurable function from $\mathbb{R}$ into $\mathbb{R}$; suppose that a noised observation of $M_n(x)$ can be made for all x.
Suppose that the zero θ_n of M_n exists and is unique ; the aim is to get an estimation of θ_n at time n.
Let H be a separable real Hilbert space with inner product $\langle\, . \,, \,.\, \rangle$. Let K = H.
Now suppose that $\theta_n = \langle \beta, U_n \rangle$ where β is an unknown element of H and U_n an element of H which is known at time n.
In order to get an estimation of β, consider the process (X_n) in H :

$$X_{n+1} = X_n - U_n a_n Y_n$$

where $a_n \in \mathbb{R}^+$; let $\Theta_n = \langle X_n, U_n \rangle$; Θ_n is an estimator of θ_n ; Y_n is a noised observation of $M_n (\Theta_n)$ such that $E [Y_n \mid F_n] = M_n (\Theta_n)$.

This process due to Ruppert [8] can be studied as a special case of the process 2.1.

2.2.4. Estimation of the vector parameter of a regression model

Notations are those of section 1. Suppose that $E [R_n \mid F_n] = 0$. Choose $H = \mathbb{R}^k$, $\forall n$, $H_n' = \mathbb{R}^{m_n}$; $\forall n$, $Z_n = 0$, $\beta_n = 0$, $Y_n = G_n (X_n) - V_n$; $\forall n$, $A_n (X_1, X_2, -, X_n ; Y_n)$

$= a_n (X_1, X_2, -, X_n).(G_n(X_n) - V_n)$. So we have the process

$$X_{n+1} = P (X_n - a_n (X_1, -, X_n).(G_n(X_n) - V_n))$$

2.2.5. Remark

In short, we can consider in the process 2.1. that $Y_n + Z_n + \beta_n = S_n$ is a noised observation at time n of $M_n (X_n)$ where M_n is a measurable function whose zero θ_n is to be estimated. If $E [S_n | F_n] = M_n (X_n)$ a.s., then we put $Y_n = S_n$, $Z_n = 0$, $\beta_n = 0$; if $E [S_n | F_n] \neq M_n (X_n)$, then we take Y_n such that $E [Y_n | F_n] = M_n (X_n)$ a.s., possibly $Y_n = M_n (X_n)$; the two types of noises Z_n and β_n will satisfy different assumptions.

3. CONVERGENCE WITH PROBABILITY ONE OF THE PROCESS

3.1. Structure of the results

The convergence is achieved by using a lemma of the martingale theory. This approach was used by Blum [2], Gladyshev [4], Robbins and Siegmund [7].

The proof of the convergence with probability one of $||X_n - \theta_n||$ to 0 has two parts. The first one is common to the three theorems given here : we use the convergence lemma of the non-negative almost supermartingales of Robbins and Siegmund [7] to prove that $||X_n - \theta_n||$ converges to a random variable T in $\mathbb{R}^+$. In the second one, we get $T = 0$ a.s. ; we show that several sets of assumptions concerning $E [A_n(X_1, X_2, -, X_n ; Y_n) | F_n]$ can be used to achieve this result.

3.2. Lemma

3.2.1. Definition of the random variables W_n

For $n \geqslant 1$, let L_n be a measurable function from H^{n+1} into H_n'. We define recursively the H valued random variables W_n :

$$W_1 = 0$$
$$W_{n+1} = W_n + A_n (X_1, X_2, -, X_n ; Z_n - L_n (X_1, -, X_n ; W_n)).$$

Thus in the case where

$\forall n$, $H_n' = H$, $L_n (X_1, X_2, -, X_n ; W_n) = W_n$, $A_n (X_1, -, X_n ; Z_n - W_n) = \frac{1}{n} (Z_n - W_n)$, we have $W_{n+1} = \frac{1}{n} \sum_1^n Z_i$; in this case also, but when $H = \mathbb{R}^k$ and $A_n (X_1, X_2, -, X_n ; Z_n - W_n) = a_n (Z_n - W_n)$ with $a_n \in \mathbb{R}^+$, Ljung gives conditions to have $W_n \xrightarrow{a.s.} 0$ [5], which is one of the assumptions concerning W_n.

3.2.2. Notations

We shall use simpler notations. $A_n (X_1, X_2, -, X_n ; Y_n + Z_n + \beta_n)$, $A_n (X_1, X_2, -, X_n ; Y_n)$, $A_n (X_1, X_2, -, X_n ; L_n (X_1, X_2, -, X_n ; W_n))$ will be denoted respectively $A_n (Y_n + Z_n + \beta_n)$, $A_n Y_n$, $A_n L_n W_n$.

3.2.3. Assumptions and statement of the lemma

We suppose in the assumptions made in this work that the sequence (θ_n) in K exists and is unique.

A. 3.2.1. $\sum_1^\infty ||\theta_n - \theta_{n+1}|| < \infty$

We shall use assumptions concerning $A_n (X_1, -, X_n ; Y_n)$.

A. 3.2.2. $\sum_1^\infty E [||A_n Y_n||^2 | F_n] < \infty$ a.s.

A. 3.2.3. $\sum_1^\infty < X_n - \theta_n, E [A_n Y_n | F_n] >^- < \infty$ a.s.

A. 3.2.2. can be replaced by

A. 3.2.2'. $\forall n, \exists D_n, E_n$, $\mathbb{R}^+$ valued random variables, F_n - measurable :

$\sum_1^\infty D_n < \infty$ a. s. ; $\sum_1^\infty E_n < \infty$ a.s. ;

$E [||A_n Y_n||^2 | F_n] \leqslant D_n ||X_n - \theta_n||^2 + E_n$ a.s.

Then we shall use assumptions concerning W_n.

A. 3.2.4. $\exists N : \forall n > N$, $\theta_n - W_n$ is a K valued random variable ;

$W_n \xrightarrow{a.s.} 0$;

$\sum_1^\infty ||A_n L_n W_n|| < \infty$ a.s.

A. 3.2.5. $\sum_1^\infty | < W_n, E [A_n Y_n | F_n] > | < \infty$ a.s.

When A.3.2.2. is replaced by A. 3.2.2'., A. 3.2.5. is replaced by

A. 3.2.5'. $\sum_1^\infty ||W_n|| \sqrt{\sup(D_n, E_n)} < \infty$ a.s.

Finally we shall use assumptions concerning β_n.

A. 3.2.6. $\sum_1^\infty E[||A_n \beta_n||^2 | F_n] < \infty$ a.s. ;

$\sum_1^\infty || E [A_n \beta_n | F_n] || < \infty$ a.s.

Lemma 3.1.

Assume A. 3.2.1. to A. 3.2.6. (A. 3.2.2. can be replaced by A. 3.2.2'. provided that A. 3.2.5. be replaced by A. 3.2.5'.). Then :

$\exists T$, $\mathbb{R}^+$ valued random variable : $||X_n - \theta_n|| \xrightarrow{a.s.} T$;

$\sum_1^\infty < X_n - \theta_n, E [A_n Y_n | F_n] >^+ < \infty$ a.s.

Proof

The detailed proof is in [6].

Using properties of projection operators gives :

$||X_{n+1} + W_{n+1} - \theta_{n+1}|| \leqslant ||X_n - A_n (Y_n + Z_n + \beta_n) - \theta_{n+1} + W_{n+1}||$

Using the recursive definition of W_n gives :

$$\|X_{n+1} + W_{n+1} - \theta_{n+1}\| \leqslant \| X_n + W_n - \theta_n - A_n Y_n - A_n L_n W_n - A_n \beta_n + \theta_n - \theta_{n+1} \|$$

It follows that : $E[\|X_{n+1} + W_{n+1} - \theta_{n+1}\|^2 \mid F_n]$

$$\leqslant \|X_n + W_n - \theta_n\|^2 - 2 < X_n + W_n - \theta_n, A_n L_n W_n + E[A_n \beta_n \mid F_n] + \theta_{n+1} - \theta_n > + 2$$

$$< X_n - \theta_n, E[A_n Y_n \mid F_n]>^- -2 < W_n, E[A_n Y_n \mid F_n]> + 4 E[\|A_n Y_n\|^2 \mid F_n] +$$

$$4\|A_n L_n W_n\|^2 + 4 E[\|A_n \beta_n\|^2 \mid F_n] + 4 \|\theta_{n+1} - \theta_n\|^2 - 2 < X_n - \theta_n,$$

$$E[A_n Y_n \mid F_n]>^+ \quad \text{a.s.}$$

We have : $|< X_n + W_n - \theta_n, A_n L_n W_n + E[A_n \beta_n \mid F_n] + \theta_{n+1} - \theta_n >|$

$$\leqslant \|X_n + W_n - \theta_n\|^2 (\|A_n L_n W_n\| + \|E[A_n \beta_n \mid F_n]\| + \|\theta_{n+1} - \theta_n\|) + \|A_n L_n W_n\|$$

$$+ \|E[A_n \beta_n \mid F_n]\| + \|\theta_{n+1} - \theta_n\| \quad \text{a.s.}$$

Now we use the Robbins-Siegmund lemma. Then : $\exists T$, $\mathbb{R}^+$ valued random variable :

$\|X_n + W_n - \theta_n\| \xrightarrow{a.s.} T$; $\sum_1^\infty < X_n - \theta_n, E[A_n Y_n \mid F_n]>^+ < \infty$ a.s. As $W_n \xrightarrow{a.s.} 0$, we have : $\|X_n - \theta_n\| \xrightarrow{a.s.} T$.

The principle of the proof is the same with the second set of assumptions.

3.3. First theorem of convergence with probability one

We shall make the following assumptions.

A. 3.3.1. $\forall n$, $\exists T_n$, a measurable function from K^n into H :

$$E[A_n Y_n \mid F_n] = T_n(X_1, X_2, -, X_n) \quad \text{a.s.}$$

A.3.3.2. $\forall\, 0 < \varepsilon < 1$, $\forall x_1, x_2, -, x_n, - \in K$,

$$\sum_1^\infty \inf_{\{x : x \in K,\ \varepsilon < \|x - \theta_n\| < \frac{1}{\varepsilon}\}} < x - \theta_n, T_n(x_1, x_2, -, x_{n-1}, x)>^+ = +\infty$$

Theorem 3.1.

If the assumptions of lemma 3.1. and A. 3.3.1, A. 3.3.2. hold, then $\|X_n - \theta_n\| \xrightarrow{a.s.} 0$.

Proof

From lemma 3.1. we have : $\|X_n - \theta_n\| \xrightarrow{a.s.} T$ and $\sum_1^\infty < X_n - \theta_n, E[A_n Y_n \mid F_n]>^+ < +\infty$ a.s. Let ω be a point of the intersection of the a.s. convergence sets and suppose $T(\omega) \neq 0$. Then by using A. 3.3.1. and A.3.3.2. we have $\sum_1^\infty < X_n(\omega) - \theta_n, E[A_n Y_n \mid F_n](\omega)>^+ = +\infty$ which is inconsistent with the second assertion of lemma 3.1. Thus $T(\omega) = 0$.

3.4. Second theorem of convergence with probability one

First we give a lemma. We shall make the following assumptions.

A.3.4.1. $\| A_n Y_n - E[A_n Y_n \mid F_n] \| \xrightarrow{a.s.} 0.$

A.3.4.2. $\| A_n \beta_n \| \xrightarrow{a.s.} 0.$

Lemma 3.2.

Assume A. 3.2.2, A. 3.2.4, A. 3.4.1, A. 3.4.2. or assume A. 3.2.1, A. 3.2.2', A.3.2.3, A. 3.2.4, A. 3.2.5', A. 3.2.6, A. 3.4.1, A. 3.4.2. Then : $\| X_{n+1} - X_n \| \xrightarrow{a.s.} 0.$

Proof

This follows from

$\| X_{n+1} - X_n \| \leqslant \| A_n Y_n - E[A_n Y_n \mid F_n] \| + \| E[A_n Y_n \mid F_n] \| + \| W_{n+1} - W_n \| + \| A_n L_n W_n \| + \| A_n \beta_n \|$ and the assumptions of the lemma.

A.3.2.1. implies for the sequence (θ_n) to be convergent ; denote $\theta = \lim_{n\to\infty} \theta_n$. We shall make the following assumptions.

A.3.4.3. $\forall n$, $\exists T_n$, a measurable function from K into H : $E[A_n Y_n \mid F_n] = T_n(X_n)$ a.s.

A.3.4.4. $\forall\, 0 < \varepsilon < 1$, $T(\varepsilon) = \sup_n \sup_{\{x : x \in K,\ \varepsilon < \|x - \theta\| < \frac{1}{\varepsilon}\}} \| T_n(x) \| < \infty.$

A.3.4.5. $\forall \varepsilon > 0$, $\exists \eta > 0$:

$$(x_1, x_2 \in K : \| x_1 - x_2 \| < \eta) \Rightarrow (\sup_n \| T_n(x_1) - T_n(x_2) \| < \varepsilon).$$

A.3.4.6. $\exists q \in \mathbb{N}$, $\exists (n_l, l \geqslant 1) \subset \mathbb{N} : n_{l+1} \geqslant n_l + q$, $\forall\, 0 < \varepsilon < 1$, $\exists L(\varepsilon) : \forall l > L(\varepsilon)$,

$$b_l(\varepsilon) = \inf_{\{x : x \in K,\ \varepsilon < \|x - \theta\| < \frac{1}{\varepsilon}\}} \sum_{j \in I_l} \langle x - \theta_j, T_j(x) \rangle^+ > 0$$

with $I_l = \{n_l, n_l + 1, \ldots, n_l + q - 1\}$; $\sum_{l=1}^{\infty} b_l(\varepsilon) = +\infty.$

Theorem 3.2.

If the assumptions of lemmas 3.1, 3.2. and A. 3.4.3. to A. 3.4.6. hold, then : $\| X_n - \theta_n \| \xrightarrow{a.s.} 0.$

Proof

From lemma 3.1, we have : $\| X_n - \theta_n \| \xrightarrow{a.s.} T$. As $\theta_n \to \theta$, we have : $\| X_n - \theta \| \xrightarrow{a.s.} T$. Let ω be a point of the intersection of the a.s. convergence sets and suppose $T(\omega) \neq 0$. Then, $\exists \varepsilon_1 > 0$: for n sufficiently large, $\varepsilon_1 < \| X_n(\omega) - \theta \| < \frac{1}{\varepsilon_1}$.

A.3.4.6. implies for l sufficiently large : $\exists m_l(\omega)$ (denoted m_l) $\in I_l$:

$\langle X_{n_l}(\omega) - \theta_{m_l}, T_{m_l}(X_{n_l}(\omega)) \rangle^+ \geqslant \frac{b_l(\varepsilon_1)}{q} > 0.$

Consider : $< X_{m_1}(\omega) - \theta_{m_1}, T_{m_1}(X_{m_1}(\omega)) > = < X_{m_1}(\omega) - X_{n_1}(\omega), T_{m_1}(X_{m_1}(\omega)) >$

$+ < X_{n_1}(\omega) - \theta_{m_1}, T_{m_1}(X_{n_1}(\omega)) > + < X_{n_1}(\omega) - \theta_{m_1}, T_{m_1}(X_{m_1}(\omega)) - T_{m_1}(X_{n_1}(\omega)) >.$

Using lemma 3.2. and A. 3.4.4. gives $< X_{m_1}(\omega) - X_{n_1}(\omega), T_{m_1}(X_{m_1}(\omega)) > \longrightarrow 0.$

Using lemma 3.2. and A. 3.4.5. gives $< X_{n_1}(\omega) - \theta_{m_1}, T_{m_1}(X_{m_1}(\omega)) - T_{m_1}(X_{n_1}(\omega)) > \longrightarrow 0.$

Its follows that for l sufficiently large, $< X_{m_1}(\omega) - \theta_{m_1}, T_{m_1}(X_{m_1}(\omega)) > \geqslant \frac{b_1(\varepsilon_1)}{2q}$.

A.3.4.6. implies $\sum_1^\infty < X_n(\omega) - \theta_n, T_n(X_n(\omega)) >^+ = +\infty$, which is inconsistent with the second assertion of lemma 3.1. Thus $T(\omega) = 0$.

3.5. Third theorem of convergence with probability one

3.5.1. A dynamic stochastic approximation case

Consider k separable real Hilbert spaces $E^1, E^2, -, E^k$ and for $i = 1, 2, -, k$, K^i a non-empty closed convex subset of E^i ; let P^i be the projection operator on K^i. For $i = 1, 2, -, k$, inner product and norm in E^i are denoted respectively $< . , . >^i$ and $\|.\|^i$.

Let $H = E^1 \times E^2 \times - \times E^k$. Inner product in H is defined by $< x , y > = \sum_{i=1}^k < x^i, y^i >^i$; norm in H is denoted $\|.\|$. Let $K = K^1 \times K^2 \times - \times K^k$; let P be the projection operator on K.

Consider for $n \geqslant 1$, H_n', a separable real Hilbert space, M_n, a measurable fuction from $\mathbb{R}^k$ into H_n', $\theta_n = (\theta_n^1, \theta_n^2, -, \theta_n^k)$, a point of $\mathbb{R}^k$ such that $M_n(\theta_n) = 0$. Suppose for $i = 1, 2, -, k$, $\theta_n^i = < \beta^i, U_n^i >^i$, where β^i is an unknown element of E^i and U_n^i an element of E^i which is known at time n.
We can have a noised observation at time n of $M_n(x)$ for all $x \in \mathbb{R}^k$. The aim is to estimate θ_n at time n. We shall use first an estimation scheme of $\beta = (\beta^1, \beta^2, -, \beta^k)$.

Consider the process (X_n^i) in K^i ; X_n^i will be an estimator of β^i.

$$X_{n+1}^i = P^i (X_n^i - U_n^i < a_n^i (\Theta_1, -, \Theta_n), Y_n + Z_n + \beta_n >_n) \qquad \text{with}$$

$\Theta_n = (\Theta_n^1, \Theta_n^2, -, \Theta_n^k)$, for $i = 1, 2, -, k$, $\Theta_n^i = < X_n^i, U_n^i >^i$, Y_n, a H_n' valued random variable such that $E[Y_n \mid F_n] = M_n(\Theta_n)$ a.s., Z_n, β_n, H_n' valued random variables, for $i = 1, 2, -, k$, a_n^i, a measurable function from $\mathbb{R}^{kn}$ into H_n', $< . , . >_n$, the inner product in H_n'. $Y_n + Z_n + \beta_n$ is a noised observation of $M_n(\Theta_n)$.

Let $X_n = (X_n^1, X_n^2, -, X_n^k)$. We have :

$$X_{n+1} = P(X_n - A_n(X_1, -, X_n ; Y_n + Z_n + \beta_n))$$

where $A_n(X_1, X_2, -, X_n ; Y_n + Z_n + \beta_n)$ is a H valued random variable whose i^{th} component is $U_n^i < a_n^i (\Theta_1, -, \Theta_n), Y_n + Z_n + \beta_n >_n$ (Θ_n depends on X_n). So the process (X_n) is a special case of the process 2.1.

3.5.2. Theorem of convergence with probability one

We apply without any difficulty lemmas 3.1. and 3.2. to this case.
The ith component of $E[A_n(X_1, X_2, -, X_n; Y_n) \mid F_n]$ is $E[U_n^i < a_n^i(\Theta_1, -, \Theta_n), Y_n >_n \mid F_n] = U_n^i < a_n^i(\Theta_1, -, \Theta_n), M_n(\Theta_n) >_n = \tau_n(\Theta_1, \Theta_2, -, \Theta_n)$ a.s.
As Θ_n depends on X_n, we have $\tau_n(\Theta_1, -, \Theta_n) = T_n(X_1, -, X_n)$.
$E[A_n(X_1, X_2, -, X_n; Y_n) \mid F_n]$ depends on $X_1, X_2, -, X_n$ (possibly on X_n only, if $a_n^1, a_n^2, -, a_n^k$ depend on Θ_n only) ; we may apply theorems 3.1. and 3.2. But it may be remarked that T_n is a compound function of $X_1, X_2, -, X_n$:

$$T_n : (X_1, -, X_n) \longrightarrow (\Theta_1, -, \Theta_n) \xrightarrow{\tau_n} \tau_n(\Theta_1, -, \Theta_n).$$

This special structure of T_n can be used to get a third theorem in the case where $E^i = \mathbb{R}^{p_i}$ for $i = 1, 2, -, k$; we suppose that we have the usual Euclidean inner product and norm in $\mathbb{R}^{p_i}$. Denote $U_n^{i'}$ the transpose of U_n^i. We shall make the following assumptions.

A.3.5.1. $\exists q \in \mathbb{N} : q \geqslant \max(p_1, p_2, -, p_k)$, $\exists (n_1, 1 \geqslant 1) : n_{1+1} \geqslant n_1 + q$, $\exists 0 < \delta < \Delta < \infty : \inf_{i,n_1} \delta_{n_1}^i \geqslant \delta$, $\sup_{i,n_1} \Delta_{n_1}^i \leqslant \Delta$, where δ_n^i and Δ_n^i are respectively the smallest and the largest eigenvalue of $\sum_{j=0}^{q-1} U_{n+j}^i \, U_{n+j}^{i\,'}$.

A.3.5.2. $\forall\, 0 < \varepsilon < 1$, $\forall x_1, x_2, -, x_j, - \in \mathbb{R}^k$,

$$\sum_1 \inf_{n_1 \leqslant j \leqslant n_1 + q - 1} \inf_{\{x : \varepsilon < \|x - \theta_j\| < \frac{1}{\varepsilon}\}} < \sum_{i=1}^{k} (x^i - \theta_j^i)\, a_j^i(x_1, -, x_{j-1}, x), M_j(x) >_j^+ = +\infty.$$

A.3.5.3. $\sup_n \|U_n\| < \infty$.

Theorem 3.3.

If the assumptions of lemmas 3.1, 3.2. and A. 3.5.1, A. 3.5.2. hold, then : $\|X_n - \beta\| \xrightarrow{a.s.} 0$. *If the assumption A. 3.5.3. holds in addition then :* $\|\Theta_n - \theta_n\| \xrightarrow{a.s.} 0$.

Proof

It is an extension of the proof given by Ruppert in the case of $k = 1$ [8].
Denote $I_1 = \{n_1, n_1+1, -, n_1 + q - 1\}$. Let ω be a point of the intersection of the a.s. convergence sets. Using A. 3.5.1. gives :

$$\sum_{p=0}^{q-1} < X_{n_1}^i(\omega) - \beta^i, U_{n_1+p}^i >^2 \geqslant \delta \|X_{n_1}^i(\omega) - \beta^i\|^2.$$

Therefore, $\exists m_1^i(\omega) \in I_1 : < X_{n_1}^i(\omega) - \beta^i, U_{m_1^i(\omega)}^i >^2 \geqslant \frac{\delta}{q} \|X_{n_1}^i(\omega) - \beta^i\|^2$ and

$$\sum_{i=1}^{k} \langle X_{n_1}^{i}(\omega) - \beta^{i}, U_{m_1(\omega)}^{i} \rangle^2 \geq \frac{\delta}{q} \| X_{n_1}(\omega) - \beta \|^2 .$$

Using lemma 3.1. gives : $\exists T$, a $\mathbb{R}^+$ valued random variable : $\| X_n - \beta \| \xrightarrow{a.s.} T$.
Suppose $T(\omega) > 0$. Then, $\exists\, B(\omega) > 0$, $\exists i_1(\omega)$ (denoted i_1) $\in \{1, 2, -, k\}$: for l sufficiently large, $| \langle X_{n_1}^{i_1}(\omega) - \beta^{i_1}, U_{m_1}^{i_1} \rangle | \geq B(\omega)$ with $m_1 = m_1^{i_1(\omega)}(\omega)$.

Consider : $\Theta_{m_1}^{i_1}(\omega) - \theta_{m_1}^{i_1} = \langle X_{m_1}^{i_1}(\omega) - X_{n_1}^{i_1}(\omega), U_{m_1}^{i_1} \rangle + \langle X_{n_1}^{i_1}(\omega) - \beta^{i_1}, U_{m_1}^{i_1} \rangle$.

Using lemma 3.2. and A. 3.5.1. gives : $\langle X_{m_1}^{i_1}(\omega) - X_{n_1}^{i_1}(\omega), U_{m_1}^{i_1} \rangle \longrightarrow 0 \ (l \to \infty)$.

Therefore, $\exists\, C(\omega) > 0$: for l sufficiently large, $|\Theta_{m_1}^{i_1}(\omega) - \theta_{m_1}^{i_1}| \geq C(\omega)$ and $\| \Theta_{m_1}(\omega) - \theta_{m_1} \| \geq C(\omega)$.

Using A. 3.5.1. gives : $\exists A(\omega) > 0$: for l sufficiently large, $\| \Theta_{m_1}(\omega) - \theta_{m_1} \| \leq A(\omega)$.
Using then A. 3.5.2. gives an assertion which is inconsistent with the second one of lemma 3.1. Thus $T(\omega) = 0$.

4. APPLICATIONS

Theorems 3.1, 3.2, 3.3. may be applied to the parameter estimation problem, specially in the case of a linear or non-linear regression model (see [6]). We give here as an example the application of theorems 3.1. and 3.2. to the estimation of the vector parameter $\theta (\theta \in \mathbb{R}^k)$ of a linear regression model, the observations being supposed uncorrelated.
Notations of section 1 are used. We consider two processes. In the first one, observed real data are grouped together in batches of size k ; we use the theorem 3.1. to prove its convergence. In the second one, only one observed real data is used at each step ; we use the theorem 3.2. to prove its convergence.
The proofs are simple and so are omitted. They can be found in [6].

4.1. First corollary

Usual Euclidean inner product and norm are used in $\mathbb{R}^k$. For any matrix A, we use as its norm $\|A\| = (\mu(A' A))^{\frac{1}{2}}$, $\mu(A' A)$ being the largest eigenvalue of $A' A$, A' being the transpose of A.
Consider the linear regression model

$$V_n = B_n \, \theta + R_n$$

where θ is a point of K, a non-empty closed convex subset of $\mathbb{R}^k$, B_n is a square matrix of order k, R_n is a $\mathbb{R}^k$ valued random variable.
We choose $a_n (X_1, X_2, -, X_n) = \frac{B_n'}{n^\alpha}$. So we have the process (X_n) :

$$X_{n+1} = P \left(X_n - \frac{B_n'}{n^\alpha} (B_n X_n - V_n)\right)$$

The following assumptions will be made.

A. 4.1.1. $\forall\, n,\ E\,[R_n \mid F_n] = 0$.

A. 4.1.2. $\frac{1}{2} < \alpha \leqslant 1$.

A. 4.1.3. $\sup_n \| B_n \| < \infty$.

A. 4.1.4. $\sup_n\ E\,[\| R_n \|^2 \mid F_n] < \infty$ a.s.

A. 4.1.5. $\sum_1^\infty \frac{\lambda_n}{n^\alpha} = +\infty$, where λ_n is the smallest eigenvalue of $B_n' B_n$.

Corollary 4.1.

Assume A. 4.1.1. to A. 4.1.5. Then : $\| X_n - \theta \| \xrightarrow{a.s.} 0$.

4.2. Second corollary

We use the notations of section 4.1. But here, B_n is a $(1 \times k)$ matrix and R_n a $\mathbb{R}$ valued random variable. The following assumptions will be made.

A. 4.2.1. $\sup_n E\,[R_n^2] < \infty$.

A. 4.2.2. $\exists\, q \in \mathbb{N} : q > k,\ \exists (n_l, l \geqslant 1) \subset \mathbb{N}$:

$n_{l+1} \geqslant n_l + q,\ \exists L$:

$\forall\, l > L,\ \lambda_l > 0$;

$\sum_1^\infty \frac{\lambda_l}{n_l^\alpha} = +\infty$, where λ_l is the smallest eigenvalue of

$\sum_{j \in I_l} B_j' B_j$, with $I_l = \{n_l, n_l + 1, -, n_l + q - 1\}$.

Corollary 4.2.

Assume A. 4.1.1. to A. 4.1.4. and A. 4.2.1, A. 4.2.2. Then : $\| X_n - \theta \| \xrightarrow{a.s.} 0$.

5. CONCLUSION

We have given three theorems of convergence with probability one of a stochastic approximation process in a non-empty closed convex subset of a Hilbert space, the observations being correlated and being not generally in the same space as the parameter to be estimated. We have shown off a structure in the results obtained by using a lemma of the martingale theory. This work includes an extension of a theorem of Ruppert concerning a dynamic stochastic approximation process ; it too extends to a more general case the study of the convergence of a stochastic approximation process with correlated observations defined by Ljung. These theorems can be applied to the parameter estimation problem, specially in the case of a regression model and thus can be added to those of Albert and Gardner.

REFERENCES

[1] Albert, A.E., Gardner, L.A., "Stochastic approximation and non-linear regression", Research Monograph 42, The M.I.T. Press, Cambridge, Massachusetts, 1967.

[2] Blum, J.R., "Multidimensional stochastic approximation methods", Ann. Math. Statist., 25, 1954, 737 - 744.

[3] Burkholder, D.L., "On a class of stochastic approximation processes", Ann. Math. Statist., 27, 1956, 1044 - 1059.

[4] Gladyshev, E.G., "On stochastic approximation", Theor. Proba. and applic., 10, 1965, 275-278.

[5] Ljung, L., "Strong convergence of a stochastic approximation algorithm", Ann. Statist., 6, 1978, 680-696.

[6] Monnez, J.M., "Etude d'un processus général multidimensionnel d'approximation stochastique sous contraintes convexes. Applications à l'estimation statistique", Thèse de Doctorat d'Etat ès Sciences Mathématiques, Université de Nancy I, 1982.

[7] Robbins, H., Siegmund, D., "A convergence theorem for non-negative almost supermartingales and some applications", Optimizing Methods in Statistics, edited by J.S. Rustagi, Academic Press, New-York, 1971, 233-257.

[8] Ruppert, D., "A new dynamic stochastic approximation procedure", Ann. Statist., 7, 1979, 1179-1195.

ON KERSTING'S THEOREM ON WEAK CONVERGENCE OF RECURSIONS

Georg Ch. Pflug

University of Gießen

Germany

1. Introduction and Assumptions.

In an interesting paper ([1]), G. Kersting considers the asymptotic distribution of a sequence of random variables $(X_n)_{n\in\mathbb{N}}$ given by the recursion

$$X_{n+1} = X_n - a_n^2 h(X_n) + a_n Y_n \qquad (1)$$

Such types of recursive schemes are encountered in the framework of Stochastic Approximation. (Kersting derives e.g. with the help of his result the asymptotic distribution of the Robbins-Monro process when the regression function is only one-sided differentiable at the root).

In this paper, Kersting's result is generalized in that way that we allow the distribution of the "error variables" Y_n is dependent on X_n. On the other hand we impose additional smoothness conditions on h and on the variances of Y_n.

Similar to G. Kersting we consider a recursive scheme of random variables of the type

$$X_{n+1} = X_n - a_n^2 h(X_n) + a_n s(X_n) Y_n \qquad (2)$$

and assume that the process $(X_n)_{n\in\mathbb{N}}$ is markovian, i.e.

$$P\{Y_n \le y \mid X_1,\dots,X_n\} = G(y|X_n) \quad \text{a.s.} \qquad (3)$$

where $G(y|x)$ is a jointly measurable family of distribution functions. We state the following set of assumptions:

(1.1) Assumptions.

(i) Let h(.) be a differentiable real function with derivative h'(.)

such that

$$h(0)=0 \qquad 0<\alpha_1 \leq |h'(\cdot)| \leq \alpha_2 < \infty$$

Furthermore we assume that for each $\eta > 0$ there is a C^∞-function h_η such that

$$\sup_x |h_\eta'(x) - h'(x)| < \eta$$

$$\text{and} \quad \limsup_n \sqrt[n]{\sup_x |h_\eta^{(n)}(x)|} < \infty$$

($h^{(n)}$ denotes the n-th derivative).

(ii) Let $s(\cdot)$ be a function satisfying a Lipschitz condition

$$|s(x)-s(y)| \leq K|x-y|$$

where $K^2 < \alpha_1$ and $0 < s_1 \leq s(x) \leq s_2$. Furthermore we assume that for each $\eta > 0$ there is a C^∞-function s_η such that

$$\sup_x |s(x)-s_\eta(x)| < -\infty$$

$$\text{and} \quad \limsup_n \sqrt[n]{\sup_x |s^{(n)}(x)|} < \infty$$

(iii) Let $(a_n)_{n \in \mathbb{N}}$ be a strictly decreasing sequence of nonnegative numbers satisfying

$$\sum a_n^2 = \infty \qquad \sum a_n^3 < \infty$$

(iv) The conditional distribution functions $G(y|x)$ satisfy

$$\int dG(y|x) = 1 \qquad \int y\,dG(y|x) = 0 \qquad \int y^2 dG(y|x) = 1$$

and

$$g(a) = \sup_x \int_{|y|>a} y^2 dG(y|x) \to 0$$

as $a \to \infty$.

2. The Theorem.

(2.1) Theorem. Let the Assumption (1.1) be fulfilled. If X_1 is a random variable with finite second moment and $(X_n)_{n \in \mathbb{N}}$ is given by recursion (2) satisfying (3) then X_n has a limiting distribution which has a density (with respect to Lebesgue measure) given by

$$\tilde{f}(x) = \frac{C}{s^2(x)} \exp\left(-2\int_0^x \frac{h(u)}{s^2(u)}\,du\right) \tag{4}$$

(C is a normalizing constant).

Proof. First of all we show that the random variables X_n have bounded second moments. By (2) we get

$$X_{n+1}^2 = X_n^2 - 2a_n^2\, h(X_n)\cdot X_n + a_n^4\, h^2(X_n) + a_n^2\, s^2(X_n)\cdot Y_n^2$$
$$+ 2a_n(X_n - a_n^2\, h(X_n))Y_n$$

By (1.1)(i) $\alpha_1 |x| \leq |h(x)| < \alpha_2 |x|$ and hence

$$E(X_{n+1}^2) < E(X_n^2)\ (1 - 2\alpha_1\, a_n^2 + \alpha_2^2\, a_n^4) + a_n^2 \cdot s_2$$

Therefore

$$E(X_n^2) < E(X_1^2) \prod_{j=1}^{n-1} (1 - 2\alpha_1 a_j^2 + \alpha_2^2 a_j^4) + s_2 \sum_{i=1}^{n-1} a_i^2 \prod_{j=i+1}^{n-1} (1-2\alpha_1 a_j^2 + \alpha_2^2 a_j^4)$$

By $\sum a_j^2 = \infty$ and $\sum a_j^3 < \infty$ the right hand side is uniformly bounded.

According to Lemma (3.2) of the appendix for every $\eta > 0$ there is a constant $B=B(\eta)$ and random variables Y_n' such that $E((Y_n' - Y_n)^2) < \eta$ and

$$E(Y_n')=0\,, \qquad E(Y_n'^2)=1 \qquad |Y_n'| < B \quad \text{a.s.}$$

The recursion (2) will now be replaced by one with more smooth functions and bounded errors. We have assumed that h' and s can be approximated by C^∞-functions. We choose an $\eta > 0$, hold it fixed and define

$$h_\eta(x) := \int_0^x h_\eta'(y)dy. \text{ Then } \sup_x |h(x) - h_\eta(x)| < \eta|x|\ .$$

Let V_o be a random variable which is independent of $X_1, Y_1, Y_2, \ldots$ and distributed according to the density $\tilde{f}_\eta$ given by

$$\tilde{f}_\eta(x) = \frac{c_\eta}{s_\eta^2(x)} \exp\left(-2 \int_0^x \frac{h(u)}{s_\eta^2(u)} du\right)$$

c_η is a normalizing constant. (A necessary enlargement of the probability space is tacitly assumed). We remark that f_η is a solution of the differential equation

$$\frac{d}{d_x}\left(s_\eta^2(x)\cdot f_\eta(x)\right) + 2h_\eta(x)\cdot f_\eta(x) = 0 \tag{9}$$

It is obvious that $\sup_x |f_\eta(x)-f(x)|$ can be made arbitrarily small by choosing η small.

We define the new recursive sequence $(V_n)_{n \geq N}$ by

$$V_{n+1} = V_n - a_n^2\, h_\eta(V_n) + a_n s_\eta(V_n)\cdot Y_n' \qquad n > N$$

$$V_N = V_o$$

The index N will be fixed later. By an argumentation completely similar to the above one can show that also the random variables V_n have uniformly bounded second moments. As we shall show X_n and V_n are closed together measured in the L^2-distance.

Utilizing the inequality $E(Y_n(Y_n - Y_n')) \leq \eta/2$ we get for $n \geq N$

$$\begin{aligned} E\,(X_{n+1} - V_{n+1})^2 &= E((X_n - V_n)^2) + a_n^4\, E\,((h(X_n) - h_\eta(V_n))^2) + \\ &+ a_n^2 E\,((s(X_n)Y_n - s_\eta(V_n)Y_n')^2) - \\ &- 2a_n^2 E\,((X_n - V_n)(h(X_n) - h_\eta(V_n))) \leq \\ &\leq E\,((X_n - V_n)^2) + 2\,a_n^4\,\alpha_2\,E\,((X_n - V_n)^2) + \\ &+ 2\,a_n^4\,\eta\,E(V_n^2) + a_n^2\,K^2\,E((X_n - V_n)^2) + \\ &+ a_n^2 s_2^2\,\eta + a_n^2\cdot\eta\cdot K - 2\,a_n^2 E((X_n - V_n)^2)(\alpha_1 - \eta) \\ &\leq E((X_n - V_n)^2)\cdot(1 - (2\alpha_1 - 2\eta - K^2)\,a_n^2 + O\,(a_n^4)) \\ &+ \eta\,O\,(a_n^2) \end{aligned}$$

Since $K^2 < \alpha_1$ we can choose η such small that $\alpha_1 - \eta - K^2 > 0$.
Hence by a well known lemma (cf. [1], Lemma (2.1)) $E((V_n - X_n)^2)$ can be made arbitrarily small, say smaller than ε by choosing an appropriate $\eta = \eta(\varepsilon)$ and a $N = N(\varepsilon)$ such that $\sum_{k=N}^{n} a_k^2 > G(\varepsilon)$ (say). Since ε is arbitrary the asymptotic distribution of V_n is arbitrarily close to that of X_n. Therefore we shall concentrate on the limiting law of $(V_n)_{n \geq N}$.

Let φ be a C^∞-function with all derivatives $\varphi^{(i)}$ bounded and let $\|\varphi\|_i = \sup_x |\varphi^{(i)}(x)|$.

By a Taylor expansion we get

$$\begin{aligned} \varphi(V_{n+1}) = \;& \varphi(V_n - a_n^2\,h_\eta(V_n)) + \\ &+ \varphi'(V_n - a_n^2 h_\eta\,(V_n))\cdot a_n\,s_\eta\,(V_n)\,Y_n' + \\ &+ 1/2\,\varphi''\,(V_n - a_n^2\,h_\eta\,(V_n))\cdot a_n^2\,s_\eta^2\,(V_n)\,Y_n'^2 + \\ &+ 1/6\,\varphi'''\,(\tilde V_n)\,a_n^3\,s_\eta^2\,(V_n)\cdot Y_n'^3 \end{aligned}$$

Where $\tilde V_n$ is an intermediate point.
Hence by the uniform boundedness of the Y_n' we obtain

$$\begin{aligned} E(\varphi(V_{n+1})) = E[\varphi(V_n) + a_n^2\,(-\varphi'(V_n)h_\eta(V_n) - \tfrac{1}{2}\,\varphi''(V_n)s_\eta^2\,(V_n)] \\ + \|\varphi\|_2\,O(a_n^4) + \|\varphi\|_3\,O(a_n^3) \end{aligned}$$

We denote by L the linear differential operator

$$(L\varphi)(x) := -\varphi'(x)\cdot h_\eta(x) + \frac{1}{2}\varphi''(x)\cdot s_\eta(x)$$

and by I the identity operator. As we just have seen

$$E(\varphi(V_{n+1})) = E((I + a_n L)\varphi(V_n)) + \|\varphi\|_2\, 0(a_n^4) + \|\varphi\|_3\, 0(a_n^3)$$

Itering this argument for $n \geq N$ we find that

$$E(\varphi(V_{n+1})) = E\Big(\prod_{k=N}^{n}(I + a_k^2 L)\,\varphi(V_o)\Big) + $$
$$+ \sum_{k=N}^{n} 0(a_k^4)\,\Big\|\prod_{l=k}^{n}(I + a_l^2 L)\,\varphi\Big\|_2 + $$
$$+ \sum_{k=N}^{n} 0(a_k^3)\,\Big\|\prod_{l=k}^{n}(I + a_l^2)\varphi\Big\|_3$$

According to (9) for any bounded C^∞-function ψ:

$$E(L\psi(V_o)) =$$
$$= \int_{-\infty}^{\infty}[-\psi'(x)h_\eta(x) + \frac{1}{2}\psi''(x)\, s_\eta^2(x)]\,\tilde{f}_\eta(x)\,dx =$$
$$= \frac{1}{2}\int_{-\infty}^{\infty}\psi(x)\,[2(h_\eta(x)\cdot\tilde{f}_\eta(x))' + (s_\eta^2(x)\,\tilde{f}_\eta(x))'']\,dx = 0$$

Hence $E\Big(\prod_{k=N}^{n}(I + a_k^2 L)\,\varphi(V_o)\Big) = E(\varphi(V_o))$

It remains to consider the rest terms. We have to look for bounds for

$$\prod_{l=N}^{n}(I + a_l^2)\,L\,\varphi = \sum_{k=1}^{n-N}\ \sum_{N\leq n_1<\ldots<n_k\leq n} a_{n_1}\cdots a_{n_k}\,(L^k\varphi)$$

Since by assumption

$$\sup_x\,(|h_\eta^{(n)}(x)| + |s_\eta^{(n)}(x)|) \leq \cdot\, H_1^n \qquad n \geq 1$$

for a constant H_1 we get (by induction) that

$$\sup_x\,|L^{(n)}\varphi(x)| \leq H_2^n$$

for a constant H_2. Therefore

$$\Big\|\prod_{k=N}^{n}(I + a_k^2 L)\,\varphi\Big\|_2 \leq \sum_{k=1}^{n-N}\ \sum_{N\leq n_1<\ldots<n_k\leq n} a_{n_1}\cdots a_{n_k}\,H_2^k$$
$$= \prod_{k=N}^{n}(1 + a_k^2\,H_2)$$

A similar inequality holds for $\Big\|\prod_{k=N}^{n}(I + a_k^2 L)\,\varphi\Big\|_3$.

If we choose now n and N large enough such that $G(\varepsilon) \leq \sum_{k=N}^{n} a_k^2 \leq G(\varepsilon) + 1$ and

$$\sum_{k=N}^{n} 0(a_k^4) \cdot \| \prod_{1=k}^{n} (I + a_1^2 L)\varphi \|_2 \le \varepsilon \qquad \text{and also}$$

$$\sum_{k=N}^{n} 0(a_k^3) \| \prod_{1=k}^{n} (I + a_1^2 L)\varphi \|_3 \le \varepsilon \qquad \text{we find that}$$

$$E(\varphi(V_{n+1})) = E(\varphi(V_o)) + 2\varepsilon \tag{10}$$

This is true for all $\varphi \in C^\infty$ with bounded derivatives, a class which uniquely determines the distribution. Hence (10) implies that the distribution of V_o is the asymptotic distribution of V_n. This concludes the proof of the theorem.

3. Appendix

(3.1) Lemma. Let Y be a square integrable random variable with expectation zero. The problem of finding a random variable Z such that

$$E((Y - Z)^2) = \min !$$

under the constraints

$$|Z| \le B \quad \text{a.s.} \qquad \text{and} \quad E(Z) = 0$$

has a unique solution given by

$$\hat{Y} = (Y + a)\Big|_{-B}^{B}$$

where $$y\Big|_{-B}^{B} = \begin{cases} B & y > B \\ y & -B \le y \le B \\ -B & y < -B \end{cases}$$ and a is chosen to satisfy $E(\hat{Y}) = 0$.

Proof. The uniqueness of the solution is a consequence at the uniform convexity of L^2. Let Z be another random variable satisfying the constraints.

Then

$$E((Y-\hat{Y})(Z - \hat{Y})) = a\, E(Z - \hat{Y}) = 0$$

Hence

$$E((Y - Z)^2) = E((Y - \hat{Y})^2) + E((\hat{Y} - Z)^2) \ge E((Y - \hat{Y})^2) .$$

(3.2) Lemma. Let $(Y_i)_{i \in I}$ be a uniformly L_2 - integrable family of random variables satisfying

$$E(Y_i) = 0 \qquad E(Y_i^2) = 1$$

Then for every $\varepsilon > 0$ one can find a $B = B(\varepsilon)$ and random variables Y_i' such that

$$E(Y_i') = 0 \qquad E(Y_i'^2) = 1$$

and $$E((Y_i - Y_i')^2) \leq 4\varepsilon .$$

Proof. Let w.l.o.g. be $\varepsilon < 1/2$. Let $B > 1 + \varepsilon$ such that

$$g := \sup_i E(Y_i^2 1_{\{|Y_i| \geq B - \varepsilon\}}) \leq \varepsilon$$

and a_i to satisfy

$$E((Y_i + a_i)|_{-B}^{B}) = 0 .$$

Then $E(|Y_i| 1_{\{|Y_i| \geq B - \varepsilon\}}) \leq \frac{g}{B-\varepsilon} \leq \varepsilon$ and

$P\{|Y_i| \geq B - \varepsilon\} \leq \frac{g}{(B-\varepsilon)^2} \leq \varepsilon$. Hence for $|a| \leq \varepsilon$

$E((Y_i + a)|_{-B}^{B}) - E(Y_i|_{-B}^{B}) \geq a(1-\varepsilon)$ and this implies

$a_i(1-\varepsilon) \leq \varepsilon$. Therefore

$$E((Y_i + a_i)|_{-B}^{B} - Y_i)^2) \leq a_i^2 + E(Y_i^2 1_{\{|Y_i| \geq B\}}) \leq$$

$$\leq \frac{\varepsilon}{(1-\varepsilon)^2} + \varepsilon < 2\varepsilon$$

Let $b_i^2 = E((Y_i + a_i)|_{-B}^{B})^2)$ and $Y_i' = \frac{1}{b_i}[(Y_i + a_i)|_{-B}^{B}]$

Then $E(Y_i') = 0$, $E(Y_i'^2) = 1$ and

$$E((Y_i - Y_i')^2) \leq 2E[((Y_i + a_i)|_{-B}^{B} - Y_i)^2] + 2(b_i - 1)^2 \leq 4\varepsilon$$

and the lemma is shown.

4. References

[1] Kersting, G.D. (1978). A weak convergence theorem with application to the Robbins-Monro process. Ann. Prob. Vol. 6, No. 6, 1015-1025

[2] Loeve, M. (1977). Probability Theory I, 4th ed. Springer Verlag New York Heidelberg Berlin.

ON CONTINUOUS TIME LEARNING MODELS

by

Helmut Pruscha

Mathematical Institute of the University Munich

1. Inter-occurrence times of learning events

Learning models (or random systems with complete connections) like the Bush-Mosteller-type models evolves on two levels, the level of (unobservable, hypothetical) states $w \in W$, $(W, \mathcal{W})$ being the state space, and on the level of (observable) events $i \in I$, where the event space

$$I = \{1,2,\ldots,m\}$$

will be supposed to be finite. The step by step transitions are governed by a transition probability P from W to I and by a measurable state transformation u from $W \times I$ to W, as illustrated in Fig.1. If a starting element $w^{(o)} \in W$ is given the model can be described by two sequences $w^{(n)}$ and ξ_n, $n \geq 1$, of random variables on a probability space $(\Omega, \mathcal{F}, \mathbb{P})$, with values in W and I, respectively (see [4] Th. 2.1.3, p. 64 and Prop. 2.1.4, p. 66).

As an example consider a linear learning model, namely the so-called linear OM-chain (Onicescu and Mihoc [8]), where W is the set of probability vectors

$$\underline{w} = (w_1, \ldots, w_m)$$

and where

$$P(\underline{w}, i) = w_i$$

$$u(\underline{w}, i)_j = \alpha_i w_j + (1-\alpha_i)\Lambda_{ij}$$

with coefficients α_i, $0 < \alpha_i < 1$ and with an m×m stochastic matrix $\underline{\Lambda} = (\Lambda_{ij})$. If all α_i are zero, we are faced with a simple Markov chain (see [9] for a detailed analysis of linear OM-chains).

Up to now events are supposed to occur on a discrete time scale. In order to implement information on the inter-occurrence times, Iosifescu [3] introduced a third sequence σ_n, $n > 0$, of $\mathbb{R}_+$-valued random variables on $(\Omega, \mathcal{F}, \mathbb{P})$, the waitimg times or interoccurrence times between two succeeding events. Now the evolution of the process is governed by a transition probability $\tilde{P}$ from W to X, where

$$X = \mathbb{R}_+ \times I,$$

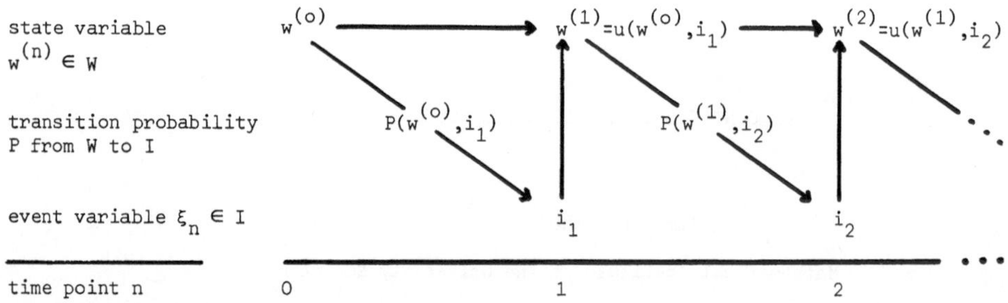

Fig. 1. Evolution of a discrete time learning model (or random system with complete connections) on the two levels of states and events, respectively.

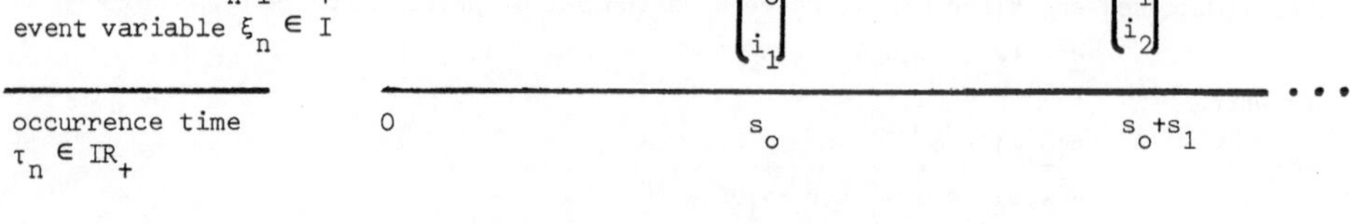

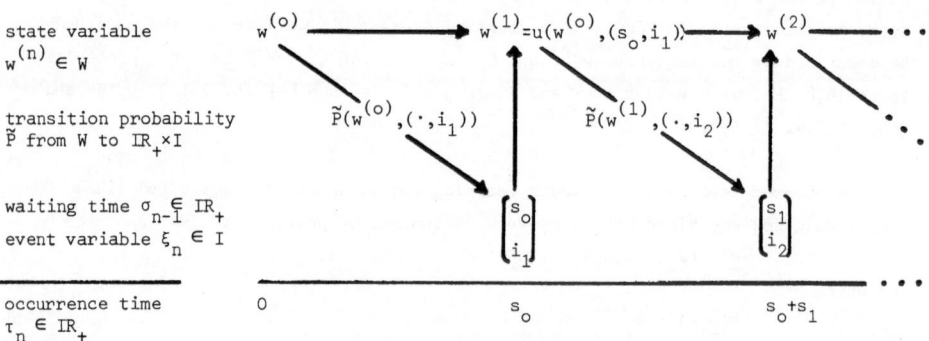

Fig.2. Evolution of a learning model (or random system with complete connections) when inter-occurrence times are implemented.

and by a measurable mapping u from $W \times X$ to W, as illustrated in Fig.2. On the basis of this approach Iosifescu established a renewal-type equation for the probabilities $\mathbb{P}(\xi_t = i), t \geq 0, i \in I$.

In order to get insight into the path-properties of such processes and to obtain a likelihood function for statistical purposes, it seems promising to base the notion of a continuous time (c.t.) learning model on the theory of multivariate point processes (see [1],[2],[5],[6]). To this end we stipulate that $\mathbb{P}$ a.e.

$$\sigma_n > 0$$
$$\sigma_n = \infty \text{ if } \sigma_{n-1} = \infty ,$$

and introduce the non-decreasing sequence of occurrence times

$$\tau_n = \sigma_o + \sigma_1 + \ldots + \sigma_{n-1} \quad (\tau_o = 0).$$

Observe that the limit

$$\tau_\infty = \lim_n \tau_n$$

can be finite (indicating an 'explosion'). On the other hand, $\tau_n = \infty$ for some $n \geq 1$ may occur (indicating an 'extinction'). We will further assume that the multivariate point process - that is the double sequence (τ_n, ξ_n), $n \geq 1$, defined on $(\Omega, F, \mathbb{P})$ - possesses an intensity. In the classical point process theory the intensity process

$$\underline{\lambda}_t = (\lambda_{1,t}, \ldots, \lambda_{m,t})$$

was introduced by

$$\lambda_{i,t} = \lim_{h \downarrow 0} (1/h) \ \mathbb{P}(N_{i,t+h} - N_{i,t} \geq 1 | \underline{N}_s, s \leq t)$$

or by

$$\lambda_{i,t} = \lim_{h \downarrow 0} (1/h) \ \mathbb{E}(N_{i,t+h} - N_{i,t} | \underline{N}_s, s \leq t),$$

where the counting process $\underline{N}_t = (N_{1,t}, \ldots, N_{m,t})$ is defined by $(i \in I, t \geq 0)$

$$N_{i,t} = \Sigma_n \, 1(\tau_n \leq t) 1(\xi_n = i)$$

Note that $N_{i,t}$ is the number of events of type i up to time t. In the martingale set-up the nonnegative process $\underline{\lambda}_t$ is called intensity if the process $\underline{N}_t - \int_o^t \underline{\lambda}_s ds$ when stopped at τ_n is a vector-valued martingale (see [2],sec. II.3). Aalen ([1],Cor.2.1, p.22) connects these two concepts of intensity.

In the sequel we will use

$$F_t = \sigma(\underline{N}_s, s \leq t)$$
$$F_{\tau_n} = \sigma((\tau_1, \xi_1), \ldots, (\tau_n, \xi_n)) = \sigma(\underline{N}_{s \wedge \tau_n}, s \geq 0)$$
$$F_\infty = \sigma((\tau_1, \xi_1), (\tau_2, \xi_2), \ldots) = \bigvee_{t \geq o} F_t .$$

(see [2],Appendix 2, sec.3 for these and other results on point process histories).

2. Regenerative intensity processes

To arrive at a definition of a continuous time learning model in terms of the intensity process $\underline{\lambda}_t$, we assume in the following that $\underline{\lambda}_t$, $t > 0$, can be defined in the following piecewise manner:

$$(1) \qquad \underline{\lambda}_t = \sum_{n \geq 0} \underline{\lambda}_t^{(n)} 1(\tau_n < t \leq \tau_{n+1}) + \underline{\lambda}_t^{(\infty)} 1(\tau_\infty < t)$$

where $\underline{\lambda}_t^{(n)}$ $[\underline{\lambda}_t^{(\infty)}]$ is nonnegative and $F_{\tau_n} \times B_+$ $[F_\infty \times B_+]$ measurable. Observe that $\underline{\lambda}_t^{(n)}$ depends on the information up to time τ_n only. If $\underline{\lambda}_t$ is a predictable process w.r.t. F_t, e.g., adapted to F_t and left-continuous, then (1) holds true cf. Jacod ([6], Lemma 3.3)

Now we introduce the following regenerative structure. Define measurable mappings

$$\underline{f}: W \times \mathbb{R}_+ \to \mathbb{R}_+^m$$
$$u: W \times X \to W$$

where $X = \mathbb{R}_+ \times I$, and assume that

$$(2) \qquad \begin{aligned} \underline{\lambda}_t^{(n)} &= \underline{f}(w^{(n)}, t-\tau_n), \quad \tau_n < t \leq \tau_{n+1}, \\ w^{(n+1)} &= u(w^{(n)}, (\sigma_n, \xi_{n+1})) \end{aligned}$$

where some $w^{(o)} \in W$ is given. Observe that $\underline{\lambda}_t^{(n)}$ has the properties stated under (1). On the interval $(\tau_n, \tau_{n+1}]$ the intensity only depends on the state $w^{(n)}$ and on the time elapsed since τ_n (see Fig.3). Through (2) the intensity is defined for $t < \tau_\infty$ only. On the interval $[\tau_\infty, \infty)$ which is non-empty in the case of explosion an extra-term as in (1) is needed.

The transition probabilities for the step by step evolution of the process will now be written down. If we set on $\{\tau_n < \infty\}$

$$(3) \qquad \begin{aligned} P(w^{(n)}, t) &= \mathbb{P}(\sigma_n \leq t \mid F_{\tau_n}) \\ P_t(w^{(n)}, i) &= \mathbb{P}(\xi_{n+1} = i \mid F_{\tau_n}, \sigma_n = t) \end{aligned}$$

we have

$$(4) \qquad \begin{aligned} P(w,t) &= 1 - \exp(-F(w,t)) \\ P_t(w,i) &= f_i(w,t)/f(w,t) \end{aligned}$$

where $f(w,t) = \Sigma_i f_i(w,t)$, $F(w,t) = \int_0^t f(w,s)ds$; see Snyder ([10], chap.5) or Brémaud ([2], p.33 and 63). Along the diagram of Fig.4 which illustrates a one-step transition we are able to simulate an outcome $(\tau_1, \xi_1), \ldots, (\tau_n, \xi_n)$ of a continuous time learning model, provided we can calculate the inverse of $F(w,t)$ as function of t. From (3) and (4) it follows for

$$\tilde{P}(w^{(n)}, (t,i)) = \mathbb{P}(\sigma_n \leq t, \xi_{n+1} = i \mid F_{\tau_n})$$

as a transition probability from W to $\bar{X}$, $\bar{X} = \bar{\mathbb{R}}_+ \times I$ that

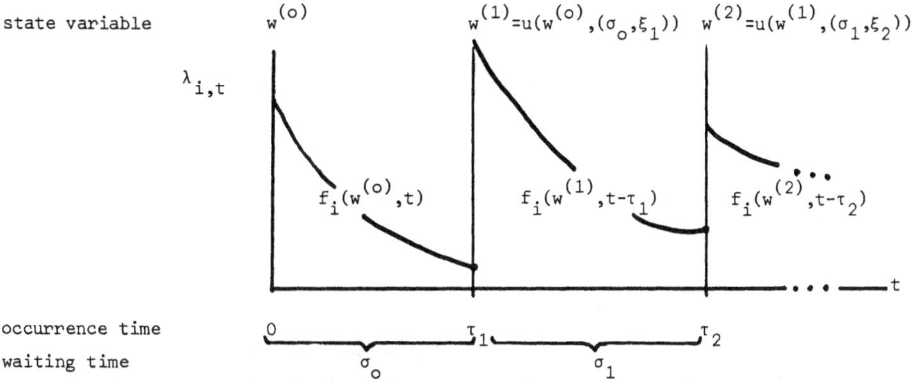

Fig.3 Sample path of the i-th component $\lambda_{i,t}$ of an intensity process with regenerative structure.

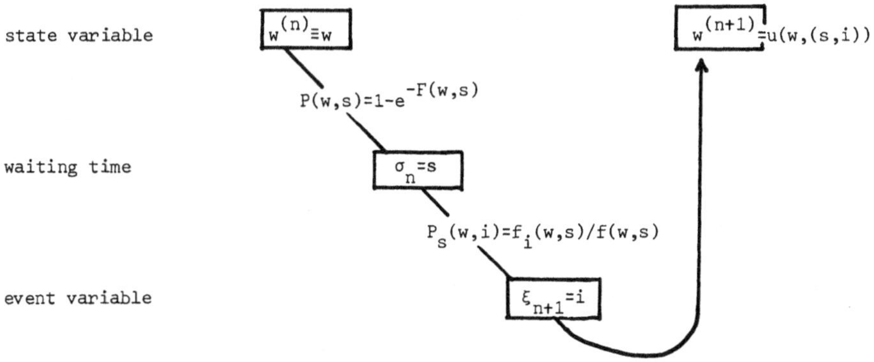

Fig.4 One step transition in a continuous time learning model (or random system with complete connections) .

$$\mathcal{P}(w,(t,i)) = \int_0^t f_i(w,s)\exp(-F(w,s))ds.$$

With respect to Fig.4 it is not surprising that the 4-tupel $(W,X,u,\mathcal{P})$ forms a random system with complete connections in such a way that the associated X-valued sequence of random variables can be identified with (σ_{n-1},ξ_n), $n \geq 1$ (see[4],p.64),provided that no extinction occurs a.e. or that the definitions of u and P are extended in an obvious way. One immediately concludes from (3) and (4) that extinction (i.e. $\sigma_n=\infty$ for some n) occurs with positive $\mathbb{P}$ probability [with $\mathbb{P}$ probability 1] if for all $w \in W$

(5) $$\lim_{t\uparrow\infty} F(w,t) < \infty \text{ [uniformly in } w\in W]$$

and that no extinction occurs $\mathbb{P}$ a.e if for all $w\in W$

(6) $$\lim_{t\uparrow\infty} F(w,t) = \infty$$

Finally, from martingale theory for point processes we get a likelihood function of a c.t. learning model. Under some mild conditions, the restricted probability measure $\mathbb{P}_t = \mathbb{P}|\mathcal{F}_t$ is absolutely continuous w.r.t. $\Pi_t = \Pi|\mathcal{F}_t$, where the probability measure Π belongs to m independent standard-Poisson processes. The Radon-Nikodym derivative $\ell_t = d\mathbb{P}_t/d\Pi_t$ is given for $\tau_n \leq t < \tau_{n+1}$ by

(7) $$\log \ell_t = \Sigma_{k=o}^{n-1} \{\log f_{\xi_{k+1}}(w^{(k)},\sigma_k) - F(w^{(k)},\sigma_k)\} - F(w^{(n)},t-\tau_n) + tm$$

see Jacod ([6], Prop. 4.3 and Th. 4.5) or Lipster and Shirayev ([7], sec. 19.4).

3. Linear learning models

We come to a continuous time linear learning model, if we set $W = \mathbb{R}_+^m$ and if $\underline{f}$ and $\underline{u}$ are linear functions in $\underline{w} = (w_1,\dots,w_m)$, e.g., $\underline{f}(\underline{w},t) = g(t)\underline{w} + c(t)$ and $\underline{u}(\underline{w},x) = a(x)\underline{w} + \underline{A}(x)$, $x = (t,i)\in X$. We become more specific and write

(8) $$\underline{f}(\underline{w},t) = \underline{w}$$
$$\underline{u}(\underline{w},x) = \alpha_i e^{-\beta t}\underline{w} + \underline{A}_i \,,$$

where $0 \leq \alpha_i \leq 1$, $\beta > 0$, $\underline{A}_i = (A_{i1},\dots,A_{im})$, $A_{ij} \geq 0$ and $\Sigma_j A_{ij}>0$ for each $i \in I$. For $\tau_n < t \leq \tau_{n+1}$ the intensity process is then given by

(9) $$\underline{\lambda}_t^{(n)} = \underline{w}^{(n)}$$

where

(10) $$\underline{w}^{(n)} = \alpha_{\xi_n} e^{-\beta\sigma_{n-1}} \underline{w}^{(n-1)} + \underline{A}_{\xi_n} \,, \quad \underline{w}^{(o)} \neq \underline{0} \text{ given.}$$

If all $\alpha_i = 0$ the c.t. linear learning model reduces to a homogeneous Markov process with infinitesimal characteristic $\underline{A} = (A_{ij})$. Iterative application of formula (1o) now brings

(11) $$w_j^{(n)} = \Sigma_{k=o}^n \eta(k+1,n)A_{\xi_k j}$$

where

$$\eta(k+1,n) = \alpha_{\xi_{k+1}} \cdots \alpha_{\xi_n} e^{-\beta(\tau_n-\tau_k)}, \qquad \eta(n+1,n) = 1$$
$$A_{\xi_o j} = w_j^{(o)}$$

It follows from (11) that $w^{(n)} \leq C(n+1)$ for some constant C such that c.t. linear learning models (8) are IP a.e. non-exploding (Jacobsen [5] , Prop. 4.4, p.44). Further no extinction takes place IP a.e. by (6), since $F(\underline{w},t) = wt$, $w = \Sigma_i w_i$.

We close by specializing the log-likelihood function (7). For $\tau_n \leq t < \tau_{n+1}$ we get for the linear model (8)

$$\text{(12)} \qquad \log \ell_t = \Sigma_{k=o}^{n-1} \{\log w_{\xi_{k+1}}^{(k)} - w^{(k)}\sigma_k\} - w^{(n)}(t-\tau_n) + mt$$

where $w^{(n)} = \Sigma_i w_i^{(n)}$. Its derivative with respect to a parameter θ is

$$\text{(13)} \qquad (\partial/\partial\theta)\log \ell_t = \Sigma_{k=o}^{n-1} \{(\partial/\partial\theta)w_{\xi_{k+1}}^{(k)} / w_{\xi_{k+1}}^{(k)} - (\partial/\partial\theta)w^{(k)}\sigma_k\} - (\partial/\partial\theta)w^{(n)}(t-\tau_n)$$

The derivatives $(\partial/\partial\theta)\underline{w}^{(k)}$ of $\underline{w}^{(k)}$ with respect to the parameters α_i, β and A_{ij} can easily be derived from (11), namely

$$(\partial/\partial\alpha_i)w_j^{(n)} = (1/\alpha_i)\, \Sigma_{k=o}^{n-1} (N_{i,\tau_n} - N_{i,\tau_k})\, \eta(k+1,n)A_{\xi_k j},$$
$$(\partial/\partial\beta)w_j^{(n)} = -\Sigma_{k=o}^{n-1} (\tau_n - \tau_k)\, \eta(k+1,n)A_{\xi_k j},$$
$$(\partial/\partial A_{ij})w_j^{(n)} = \Sigma_{k=o}^{n} 1(\xi_k = i)\, \eta(k+1,n), \qquad (\partial/\partial A_{ij})w_{j'}^{(n)} = 0 \text{ for } j \neq j'.$$

References

[1] O.O.Aalen (1976). Statistical inference for a family of counting processes. University of Copenhagen, Institute of Math.Statist.

[2] P.Brémaud (1981). Point Processes and Queues. Springer-Verlag, N.Y.

[3] M.Iosifescu (1968). Processus aléatoires à liaisons complètes purement discontinus. C.R.Acad.Sci.Paris 266, A, 1159-61.

[4] M.Iosifescu and R.Theodorescu (1969). Random Processes and Learning. Springer-Verlag, N.Y.

[5] H.Jacobsen (198o). Lecture notes on counting processes. University of Copenhagen, Institute of Math.Statist.

[6] J.Jacod (1975). Multivariate point processes. Z.Wahrscheinlichkeitsth. 31, 235-53.

[7] R.S.Lipster and A.N.Shiyayev (1978). Statistics of Random Processes. Volume II. Springer-Verlag, N.Y.

[8] O.Onicescu and G.Mihoc (1935). Sur les chaînes de variables statistiques. Bull. Soc.Math.France 59, 174-92.

[9] H.Pruscha and R.Theodorescu (1981). On a non-Markovian model with linear transition rule. Coll.Math. 44, 165-73.

[1o] D.L.Snyder (1975). Random Point Processes. Wiley, N.Y.

CONVERGENCE OF STOCHASTIC APPROXIMATION ALGORITHMS WITH NON-ADDITIVE DEPENDENT DISTURBANCES AND APPLICATIONS

by

David Ruppert
Department of Statistics
University of North Carolina
Chapel Hill, NC 27514 USA

1. Introduction. This paper is concerned with the strong convergence of recursive estimators which are generalizations of the Robbins-Monro (1951) stochastic approximation procedure. The Robbins-Monro procedure and its generalizations have been investigated in many contexts, for example recursive nonlinear regression (Albert and Gardner, 1967), recursive maximum likelihood estimation (Fabian, 1978), robust estimation of parameters for autoregressive process (Campbell, 1982), robust estimation of a location parameter (Martin and Masreliez, 1975 and Holst, 1980, 1982), control of physical processes (Comer, 1964 and Ruppert, 1979, 1981), and system identification (Kushner and Clark, 1978, section 2.6). In this paper, we present a rather general convergence theorem which allows the disturbances to be dependent and enter in a non-additive fashion. As examples, the theorem is applied to the nonlinear regression estimator of Albert and Gardner (1967), and to the author's (Ruppert 1979, 1981) Robbins-Monro type procedures for use where the root of the unknown regression function varies with time.

The algorithm studied is

$$x_{n+1} = x_n - n^{-1} H_n\{h(x_n,\xi_n)+v_n\}$$

where x_n and v_n are random vectors in $\mathbb{R}^p$, H_n is a positive definite matrix, and ξ_n is a random element in a metric space M. The standard theory of stochastic approximation applies directly when $\xi_1,\xi_2,\ldots$ are i.i.d. and when $\{v_n\}$ is a martingale difference sequence. In this paper, we still apply a standard result (Derman and Sacks, 1959), but the application is not quite so direct. In particular, the lemma of Derman and Sacks is applied not to x_n but to a subsequence $x_{n(k)}$ where $(n(k+1)-n(k)) \to \infty$ as $k \to \infty$. To complete the convergence proof, we then show that

$$\sup\{\| x_\ell - x_{n(k)} \|: \ n(k)+1 \le \ell \le n(k+1)\} \to 0 \text{ as } k \to \infty .$$

To verify the assumptions of Derman and Sacks's lemma, we utilize a result by Ranga Rao (1962) on the relationship between weak and uniform convergence of probability measures.

An alternate approach to convergence is given in Kushner and Clark (1978). They need to assume that the stochastic approximation process is bounded almost surely, but in some applications showing boundedness is nearly as difficult as proving convergence. A major advantage to using uniform convergence arguments is that boundedness need not be assumed. However, as we will point out in this paper, assuming

---

This research was supported by the National Science Foundation through Grants MCS78-01240 and MCS81-00748.

boundedness of the process does allow us to weaken other assumptions.

2. Notation and assumptions. All random variables are defined on a probability space (Ω,F,P) and all relations between random variables are meant to hold with probability 1. Let $(\mathbb{R}^k,B^k)$ be k-dimensional Euclidean space with the Borel σ-algebra, and let (M,d) be a separable metric space with the Borel σ-algebra F_M. All functions which we consider between metric spaces are assumed to be Borel measurable. Let a prime denote matrix transposition. For a real matrix A, $\|A\|=(\text{Trace } A'A)^{\frac{1}{2}}$. We will need the following assumptions, which are discussed below.

A1. $h(\cdot,\cdot)$, $h_1(\cdot,\cdot)$, and $h_2(\cdot)$ are functions from $\mathbb{R}^p\times M$ to $\mathbb{R}^p$, $\mathbb{R}^p\times M$ to $\mathbb{R}^{p\times r}$, and $\mathbb{R}^p$ to $\mathbb{R}^r$ respectively such that

$$h(x,\xi) = h_1(x,\xi)h_2(x) \ .$$

A2. For each n, H_n is a p×p positive definite symmetric random matrix. For positive random variables $\underline{\lambda} \le \overline{\lambda}$, all eigenvalues of H_n are between $\underline{\lambda}$ and $\overline{\lambda}$ for all n.

A3. Suppose μ is a probability measure on (M,F_M).

A4. Let $2 \le r \le \infty$. Let $\xi_1,\xi_2,\ldots$ be a sequence of random vectors in M. Suppose that $\int g\, d\mu = 0$ and $g \in L^r(\mu)$ implies that there exists c such that

$$\text{(2.1)} \qquad E \max_{m\le \ell\le n} \Big(\sum_{i=m}^{\ell} c_i g(\xi_i)\Big)^2 \le c \sum_{i=m}^{n} c_i^2$$

for all $n > m$ and all constants $c_m,\ldots,c_n$.

A5. There exists a nonnegative continuous function h_3 on M such that (i) $h_3\in L^r(\mu)$ and (ii) $\|h_1(x,\xi)\| \le h_3(\xi)$ for all x and ξ.

A6. $\{h_1(x,\cdot)\}_{x\in\mathbb{R}^p}$ is an equicontinuous family on M, that is, for each $\xi \in M$ and $\varepsilon > 0$ there exists $\delta > 0$ such that $\xi^* \in M$ and $d(\xi^*,\xi) < \delta$ implies that $\|h_1(x,\xi) - h_1(x,\xi^*)\| < \varepsilon$ for all $x \in \mathbb{R}^p$.

Remark. The decomposition of h into h_1 and h_2 allows some flexibility in the application of the result of Ranga Rao (1962) on uniform convergence.

If h satisfies the assumptions below on h_1, then we may simply choose $h_1 \equiv h$, $h_2 \equiv 1$, and $r = 1$. In section 4, we show an example where h does not satisfy these assumptions and therefore $h \not\equiv h_1$ and $h_2 \not\equiv 1$. In any application, there may be many suitable choices of h_1 and h_2. Note that h_1 and h_2 are used only in the *proof* of convergence; they are not employed in the algorithm itself.

Notation. Define $\overline{h}_1(x) = \int h_1(x,\xi)d\mu(\xi)$ and $\overline{h}(x) = \overline{h}_1(x)h_2(x) = \int h(x,\xi)d\mu(\xi)$.

A7. h_2 is bounded in a neighborhood of 0.

A8. For all $\varepsilon > 0$, $\inf_{\|x\|>\varepsilon} \min\{\|\overline{h}_1(x)\|, \|\overline{h}_2(x)\|\} > 0$.

A9. Suppose that $\overline{h}$ is the gradient of V, (i) $\inf_{\|x\|>\varepsilon} (V(x)-V(0)) > 0$ for all $\varepsilon>0$, and (ii) with $\ddot{V}$ the Hessian of V, $\|\ddot{V}(x)\| \le K$ for some K and all x.

Remark. The assumption of a bounded Hessian is common in the literature of multidimensional stochastic approximation. See e.g. Fabian (1971).

A10. There exists a nonnegative function h_4 on $\mathbb{R}^q$ such that $h_4 \in L^r(\mu)$, and for all ξ,x, and x',

$$\|h(x,\xi)-h(x',\xi)\| \le h_4(\xi)\|x-x'\| .$$

Notation. Fix $\alpha > 2$, define n(k) to be the integer part of k^α, and define $\rho_k = \sum_{i=n(k)}^{n(k+1)-1} i^{-1}$.

A11. $\sup_{n(k)\le\ell\le n(k+1)-1} \|H_\ell - H_{n(k)}\| = o(1)$ as $k \to \infty$.

A12. $\{v_n\}$ is a sequence of random vectors such that

$$\sup_{n(k)\le\ell\le n(k+1)-1} \|\sum_{i=n(k)}^{\ell} i^{-1} v_i\| = o(\rho_k) \text{ as } k \to \infty .$$

Remark. If $\xi_1,\xi_2,\ldots$ is a random sequence with stationary marginal distribution μ, then we can verify A4 in the case of independence using Kolmogorov's inequality, and for weak dependence by a theorem of McLeish (1975, theorem 1.6). McLeish's results can also be useful in the verification of A12.

For example, by results in section 2 of McLeish, A4 holds if the sequence $\{\xi_n\}$ is strictly stationary with marginal distribution μ and if it possesses ϕ-mixing constants which are of size $-r/(2r-2)$ or, if $r > 2$ and it possesses strong-mixing constants which are of size $-r/(r-2)$. The concept of "size" is defined in section 1 of McLeish. There it is noted that $\psi_n \ge 0$ and $\sum_{n=1}^\infty \psi_n^\theta < \infty$ imply that ψ_n is of size $-q$ for all $q < 1/\theta$. Under rather general conditions, A4 holds if $\{\xi_n\}$ is a vector-valued autoregressive, moving average process (ARMA). This follows from a paper by Pham and Tran (1980), which shows that under weak conditions an ARMA process is strong-mixing, and the mixing constants go to zero exponentially fast.

3. General results.

Lemma 3.1. *Suppose* A3 *and* A4 *hold. For* ℓ *in* $\{n(k),\ldots, n(k+1)-1\}$, *define the random probability measure* $\mu_{k,\ell}$ *by*

$$\mu_{k,\ell}(A) = \rho_k^{-1}\{\sum_{i=n(k)}^{\ell} i^{-1} I(\xi_i \in A) + \sum_{i=\ell+1}^{n(k+1)-1} i^{-1} \mu(A)\}$$

for A *in* F_M . *(Here* I(B) *is the indicator function of set* B*). Then,* $\mu_{k,\ell}$ *converges weakly to* μ *as* $n(k) + \ell \to \infty$. *(The indices* (k,ℓ) *can be ordered into a sequence according to the magnitude of* $n(k) + \ell$*).*

Proof. Suppose g is bounded (or more generally in $L^r(\mu)$). Then by A4,

$$E \max_{n(k)\le\ell\le n(k+1)-1} (\int g d\mu_{k,\ell} - \int g d\mu)^2 = O(\rho_k^{-2} \sum_{\ell=n(k)}^{n(k+1)-1} \ell^{-2}) = O(k^{1-\alpha}) .$$

Thus since $\alpha > 2$,

$$\sum_{k=1}^\infty E \max_{n(k)\le\ell\le n(k+1)-1} (\int g d\mu_{k,\ell} - \int g d\mu)^2 < \infty ,$$

whence $\max_{n(k)\le\ell\le n(k+1)-1} (\int g d\mu_{k,\ell} - \int g d\mu) \to 0$ as $k \to \infty$.

By Theorem 6.6 of Parthasarathy (1967), there exists a sequence of bounded functions

$g_1, g_2, \ldots$ such that, for any measures $\{\nu_n\}_{n=1}^{\infty}$ and ν on (M, F_M) we have $\nu_n \to \nu$ weakly if and only if $\int g_\ell \, d\nu_n \to \int g_\ell \, d\nu$ as $n \to \infty$ for each ℓ. The lemma follows. □

Lemma 3.2. *As* $k \to \infty$,

$$\sup_{x \in \mathbb{R}^p} \max_{n(k) \le \ell \le n(k+1)-1} \rho_k^{-1} \left| \sum_{i=n(k)}^{\ell} \ell^{-1} (h_1(x, \xi_\ell) - \bar{h}_1(x)) \right| \to 0 .$$

Proof. The lemma follows from A4, A5, A6, Lemma 3.1, and Theorem 3.2 of Ranga Rao (1962).

Theorem 3.1. *Under* A1 *to* A12, $x_n \to 0$ *where* x_n *is defined by the recursion*

$$x_{n+1} = x_n - n^{-1} H_n \{h(x_n, \xi_n) + v_n\}$$

Remark. As can be seen in the proof, if $\sup_n \|x_n\| < \infty$, then A8 and A9(i) can be weakened to

A8*. For all $\varepsilon > 0$, $\inf_{\varepsilon^{-1} > \|x\| > \varepsilon} \min\{\|\bar{h}_1(x)\|, \|\bar{h}_2(x)\|\} > 0$.

A9(i)*. For all $\varepsilon > 0$, $\inf_{\varepsilon^{-1} > \|x\| > \varepsilon} (V(x) - V(0)) > 0$.

Proof of Theorem 3.1. For $n(k) + 1 \le \ell \le n(k+1) - 1$,

$$\begin{aligned} x_{\ell+1} &= x_{n(k)} - \textstyle\sum_{i=n(k)}^{\ell} i^{-1} H_i \, \bar{h}(x_{n(k)}) \\ &\quad - \textstyle\sum_{i=n(k)}^{\ell} i^{-1} H_{n(k)} (h(x_{n(k)}, \xi_i) - \bar{h}(x_{n(k)})) \\ &\quad - \textstyle\sum_{i=n(k)}^{\ell} i^{-1} (H_i - H_{n(k)}) (h(x_{n(k)}, \xi_i) - \bar{h}(x_{n(k)})) \\ &\quad - \textstyle\sum_{i=n(k)}^{\ell} i^{-1} H_i (h(x_i, \xi_i) - h(x_{n(k)}, \xi_i)) - \sum_{i=n(k)}^{\ell} i^{-1} H_i v_i \\ &= x_{n(k)} - R_{k,\ell} - S_{k,\ell} - T_{k,\ell} - U_{k,\ell} - V_{k,\ell}, \text{ say.} \end{aligned} \tag{3.1}$$

By A2,

$$\|R_{k,\ell}\| \le \bar{\lambda} \rho_k \|\bar{h}(x_{n(k)})\|, \text{ and} \tag{3.2}$$

$$\|R_{k,n(k+1)}\| \ge \underline{\lambda} \rho_k \|\bar{h}(x_{n(k)})\| . \tag{3.3}$$

By A2 and Lemma 3.2,

$$\|S_{k,\ell}\| = o(\rho_k \|\bar{h}_2(x_{n(k)})\|) . \tag{3.4}$$

By A11,

$$\|T_{k,\ell}\| = o\left(\sum_{i=n(k)}^{n(k+1)-1} i^{-1} (\|h(x_{n(k)}, \xi_i)\| + \|\bar{h}(x_{n(k)})\|)\right) .$$

Moreover, by A1, A4, and A5,

$$\begin{aligned} &\textstyle\sum_{i=n(k)}^{n(k+1)-1} i^{-1} \|h(x_{n(k)}, \xi_i)\| \\ &\le \|h_2(x_{n(k)})\| \textstyle\sum_{i=n(k)}^{n(k+1)-1} i^{-1} h_3(\xi_i) = O(\rho_k \|h_2(x_{n(k)})\|) , \end{aligned}$$

and

$$\sum_{i=n(k)}^{n(k+1)-1} i^{-1} \|\bar{h}(x_{n(k)})\| = O(\rho_k \|\bar{h}_2(x_{n(k)})\|) .$$

Therefore,

$$\|T_{k,\ell}\| = o(\rho_k \|\bar{h}_2(x_{n(k)})\|) . \tag{3.5}$$

Also, by A2 and A12 we have

$$(3.6) \qquad \| V_{k,\ell} \| = o(\rho_k) \ .$$

Next, by A2, A4, and A10,

$$(3.7) \qquad \| U_{k,\ell} \| = O(\rho_k \max_{n(k) \le i \le \ell-1} \| x_i - x_{n(k)} \|) \ .$$

By (3.1), (3.2), and (3.4)-(3.7),

$$\| x_\ell - x_{n(k)} \| \le M_k \rho_k (\| \bar{h}_2(x_{n(k)}) \| + \max_{n(k) \le i \le \ell-1} \| x_i - x_{n(k)} \| + 1)$$

where $M_k = O(1)$. Then by induction,

$$(3.8) \qquad \max_{n(k) \le \ell \le n(k+1)-1} \| x_\ell - x_{n(k)} \| \le \frac{M_k \rho_k (\| \bar{h}_2(x_{n(k)}) \| + 1)}{1 - M_k \rho_k}$$

for all k so large that $M_k \rho_k < 1$. By (3.7) and (3.8),

$$\| U_{k,\ell} \| = O[(\rho_k^2 (\| \bar{h}_2(x_{n(k)}) \| + 1)] \ .$$

Now (3.1), (3.4), (3.5), and (3.8) imply that

$$x_{n(k+1)} = x_{n(k)} - \sum_{i=n(k)}^{n(k+1)-1} i^{-1} H_i \bar{h}(x_{n(k)}) + o(\rho_k \| \bar{h}_2(x_{n(k)}) \|) + o(\rho_k) \ .$$

By A5, $\bar{h}_1(x)$ is bounded, so by A9, there exist $\varepsilon_k \downarrow 0$ such that

$$V(x_{n(k+1)}) \le V(x_{n(k)}) - \underline{\lambda} \rho_k \| \bar{h}(x_{n(k)}) \|^2 + \varepsilon_k \rho_k (\| \bar{h}_2(x_{n(k)}) \|^2 + 1) \ .$$

Now choose $\alpha_k \downarrow 0$ such that $\varepsilon_k / \alpha_k^2 \to 0$ and $\sum \rho_k \alpha_k^2 = \infty$, then choose $\beta_k \downarrow 0$ such that $\|x\| > \beta_k$ implies that $\| h_1(x) \|^2 > \alpha_k$ and $\| \bar{h}(x) \|^2 > \alpha_k$, and finally find $\gamma_k \downarrow 0$ such that $\|x\| \le \beta_k$ implies that $V(x) - \underline{\lambda} \rho_k \| \bar{h}(x) \|^2 + \varepsilon_k \rho_k (\| \bar{h}_2(x_{n(k)}) \|^2 + 1) \le \gamma_k + V(0)$. This can be done since $\rho_k \approx \alpha n^{-1}$, and by A7 and A8. Then, for k sufficiently large,

$$V(x_{n(k+1)}) - V(0) \le \max\{\gamma_k, V(x_{n(k)}) - V(0) - \underline{\lambda} \rho_k \alpha_k^2 / 2\} \ .$$

Lemma 1 of Derman and Sacks (1959) implies that $V(x_{n(k)}) \to V(0)$, and then $x_{n(k)} \to 0$ by A9(i), whence $x_n \to 0$ by (3.8). □

4. Application to nonlinear regression. Albert and Gardner (1967) investigated nonlinear regression problems where, for n = 1,2,..., one observes Y_n such that for a known function F_n, and unknown vector parameter θ, and a mean-zero random variable e_n,

$$(4.1) \qquad Y_n = F_n(\theta) + e_n \ .$$

They considered estimators of θ defined by the recursion

$$(4.2) \qquad \hat{\theta}_{n+1} = \hat{\theta}_n + a_n [Y_n - F_n(\hat{\theta}_n)] \ ,$$

where a_n is a suitably chosen vector. Although it is, of course, possible to use nonlinear least squares methodology here, Albert and Gardner were interested in situations where the Y_n's are observed sequentially, and one needs to rapidly update one's estimate as each new observation arrives. Besides its use for "on-line" estimation, this recursive nonlinear estimator may be useful when handling large data sets and models with large numbers of parameters. Then, because of its recursive nature, the calculation of the estimator has modest storage requirements.

In their study of "optimal" values of a_n in (4.2), Albert and Gardner used a Taylor series linearization and their calculation of the "optimal" a_n in the linear case and were led to the choice:

$$a_n = B_n b_n \tag{4.3}$$

where $b_n = \dot{F}_n(\eta_n)$ ($\dot{F}_n$ is the gradient of F_n),

$$B_n = (B_0^{-1} + \textstyle\sum_{j=1}^n b_j b_j')^{-1} , \tag{4.4}$$

and η_n is either θ_0, a guessed value of θ, or $\eta_n = \hat{\theta}_m$ for some $m \le n$. Also B_n can be calculated recursively and without matrix inversions, except possibly for B_0; see their equation (7.45).

Albert and Gardner do not actually prove that the algorithm converges to θ for this value of a_n, but instead they analyze a different algorithm. Let P be a convex set and let $[\ ]_P$ denote the operation of projection into P. Then, they find sufficient conditions for $\tilde{\theta}_n$ defined by

$$\tilde{\theta}_{n+1} = [\tilde{\theta}_n + a_n[Y_n - F_n(\tilde{\theta}_n)]]_P$$

to converge when a_n is given by (4.3) One of these conditions is troublesome: P must lie within the ball of radius R centered at θ. The value of R is not given but can be found by examining their proof. When R is small, one will need good prior knowledge of θ.

Also, the need to project into P complicates the algorithm, perhaps unnecessarily in some applications.

In this paper, we will suppose that $F_n(\theta) = F(Z_n,\theta)$, where F is a function from $\mathbb{R}^{q-1} \times \mathbb{R}^p$ to $\mathbb{R}$ and Z_n is a known vector. However, we will not require any prior knowledge of θ. Also, we will not require that $e_1, e_2, \ldots,$ be independent (or even uncorrelated). Since recursive estimation is often used when the data form a time series, correlated errors should be allowed.

<u>B1</u>. Suppose (4.1) holds with $F_n(\theta) = F(Z_n,\theta)$, where F is a known function on $\mathbb{R}^{q-1} \times \mathbb{R}^p$ and the "independent variable," Z_n, is a known element of $\mathbb{R}^{q-1}$.

<u>B2</u>. Suppose $B(Z,\theta)$ is the gradient of $F(Z,\theta)$ with respect to θ and $b_n(\theta) = b(Z_n,\theta)$. Let H_n be a sequence of positive definite, symmetric matrices satisfying A2 and A11. Suppose (1.2) holds with $a_n = n^{-1}H_n b_n$.

<u>B3</u>. Let $\mu = \mu_1 \times \mu_2$, where μ_1 and μ_2 are probability measures on $\mathbb{R}^{q-1}$ and $\mathbb{R}$ respectively. Define $\xi_n' = (z_n', e_n)$. Assume that for some $r \in [2,\infty]$ μ and $\xi_1, \xi_2, \ldots$ satisfy A4 with $(M, F_M) = (\mathbb{R}^q, B^q)$.

Notation. Define

$$h(z,e,x) = (F(z,x) - F(z,\theta) - e)b(z,x) ,$$

$$V(x) = \tfrac{1}{2}\int (F(z,x) - F(z,\theta) - e)^2 \, d\mu(z,e) \quad , \quad \text{and} \tag{4.5}$$

$$\bar{h}(x) = \int h(z,e,x)\,d\mu(z,e) \ .$$

B4. Assume that $\bar{h}(x)$ is the gradient of $V(x)$, i.e. that the RHS of (4.5) can be differentiated under the integral sign. Also assume that h, V, and μ satisfy A1 and A5 to A10.

Remarks. It is desirable to know when A2 and A11 hold if $H_n = nB_n$ and B_n is defined by (4.4). Define $H(x) = \int b(z,x)(b(z,x))'\,d\mu_1(z)$. Suppose there exists positive constants $\underline{\lambda}$ and $\bar{\lambda}$ such that, for all x, all eigenvalues of $H(x)$ lie between $\underline{\lambda}$ and $\bar{\lambda}$. Suppose, also, that $\|b(z,x)\|^2 \le h_6(z)$, where $h_6 \in L^r(\mu_1)$ and h_6 is continuous, that $\{b(\cdot,x)(b(\cdot,x))' \colon x \in \mathbb{R}^p\}$ is an equicontinuous family on $\mathbb{R}^{q-1}$, and η_ℓ (as in equation (4.4)) is equal to $x_{n(k)}$ if ℓ is in $\{n(k),\ldots, n(k+1)-1\}$. Then, using a proof like that of Lemma 3.2, one can show that A2 and A11 hold. In the special case of linear regression, $b(z,x)$ does not depend upon x, so the value of η_ℓ is not relevant.

When z_n is a nondegenerate random vector, then B3 essentially implies that z_n and e_n are independent for each n, a condition typically used in regression analysis. Also, A4 can often be verified using the results mentioned at the end of Section 2. If the z_n are degenerate random vectors, then the next lemma may be useful in the verification of A4. It covers the situation where the errors are i.i.d. and either (i) the independent variables are selected by repeating some finite design or (ii) the Y_n form a time series with a periodic mean function.

Lemma 4.1. *Let* M *be a positive integer and let* $\alpha_1,\ldots,\alpha_M$ *be elements of* $\mathbb{R}^{q-1}$. *Suppose* $z_n = \alpha_j$ *if* n = j modulo M. *Let* $\{e_n\}_{n=1}^{\infty}$ *be i.i.d. random variables. Define the probability measure* μ *on* $\mathbb{R}^q$ *by*

$$\mu(A) = \frac{1}{M}\sum_{j=1}^{M} P((\alpha_j', e_1)' \in A) \ ,$$

i.e. μ *is the product of the marginal distribution of* e_1 *with counting measure on* $\{\alpha_1,\ldots,\alpha_M\}$. *Suppose* $Eg^2(\alpha_j, e_1) < \infty$ *for some function* g *from* $\mathbb{R}^q$ *to* $\mathbb{R}$ *and* j=1,...M. *Then* $\xi_n = (z_n', e_n)$ *and* μ *satisfy* (2.1).

Proof. The proof uses Kolmogorov's inequality and straightforward algebra. □

Remark. The requirement that $\{e_n\}$ be i.i.d. could be weakened considerably, but we will not pursue this matter here.

Theorem 4.2. *Under* B1 *to* B4, $\hat{\theta} \to \theta$.

Proof. Without loss of generality, we may take $\theta = 0$. Then the theorem follows from Theorem 3.1. Use $v_n \equiv 0$. □

Example (linear regression). Suppose $Y_n = z_n'\theta + e_n$ and $\xi_n' = (z_n', e_n)$ satisfies B3. Suppose $\int (e^r + \|z\|^{2r})\,d\mu(z,e) < \infty$, and define $\Sigma_\mu = \int zz'\,d\mu(z,e)$. Then let $r = 2$,

$$V(\theta) = \tfrac{1}{2}(x-\theta)'\,\Sigma_\mu(x-\theta) \ , \ h(z,e,x) = (z'(x-\theta)+e)z \ , \ \bar{h}(x) = \Sigma_\mu(x-\theta) \ ,$$

$$h_1(z,e,x) = [\min\{\|x-\theta\|^{-1},1\}zz'(x-\theta) \quad ze] \ ,$$

$$h_2(x) = [\max(\|x-\theta\|,1) \quad 1]' \ ,$$

$$h_3(z,e) = (\|z\|^4 + \|z\|^2 e^2)^{\frac{1}{2}}, \quad h_4(z,e) = \|z\|^2, \text{ and } B_n = \sum_{j=1}^n z_j z_j'.$$

One may check that Theorem 4.2 can be applied here. One needs to assume that B3 holds.

5. Application: A Robbins-Monro process where the unknown root changes with time. Here we first give a brief introduction to a procedure which is motivated and described in more detail in Ruppert (1979 and 1981). After that, we outline how to weaken the convergence criteria in Ruppert (1981).

Suppose that the output of some physical process can be modeled as follows. The process is influenced by an exogeneous variable s taking values in a abstract space S, and by a real valued control variable x. The variable s can be measured, but not controlled by the experimenter. The output of the process is a random variable with expectation, conditional upon s and x, equal to $M(x,s)$ for some function M. It is assumed that there exists a real valued function f of s such that $M(f(s),s) \equiv 0$, and that it is desired that the output of the process be kept as close to 0 as possible. Therefore, one should choose $x = f(s)$. However, initially neither M nor f is known. It is further assumed that $f(s) = \beta_0' U(s)$ where U is a known function from S to $\mathbb{R}^p$ and β is an unknown element of $\mathbb{R}^p$. Therefore, in order to control the process one should estimate β_0.

Ruppert (1979, 1981) studied the following recursive method of estimating β_0. Let $\hat{\beta}_n$ be the estimate of β_0 after (n-1) runs of the process. Let s_n be the value of s (assumed known) at the beginning of the nth run, and define $U_n = U(s_n)$. Define $\hat{\beta}_{n+1}$ by

$$\hat{\beta}_{n+1} = \hat{\beta}_n - n^{-1} h(s_n)\{M(\hat{\beta}_n' U_n, s_n) + e_n\} DU_n$$

where D is a symmetric, positive definite matrix, h is a positive function, and e_n is a random variable. In practice, D and h are chosen by the statistician, and the choices which maximize the speed of convergence of $\hat{\beta}_n$ to β_0 are discussed in Ruppert (1981). The variable e_n is the difference between +he output of the process on the nth run and $M(\hat{\beta}_n' U_n, s_n)$.

Now to apply theorem 3.1 we set $x_n = \hat{\beta}_n$, $H_n = D$, $\xi_n = s_n$, $v_n = h(s_n)e_n$, $\mathcal{M} = S$ and $h(x,\xi) = h(\xi)\ M(x'U(\xi),\xi)\ DU(\xi)$.

Application of theorem 3.1 is then completely straightforward. In an expanded version of this paper, which is available from the author, the details are given. The assumptions are considerably weaker than in Ruppert (1981).

REFERENCES

Albert, A.E. and Gardner, L.A., Jr. (1967). *Stochastic Approximation and Nonlinear Regression.* The M.I.T. Press, Cambridge, Mass.

Campbell, Katherine. (1982). Recursive computation of M-estimates for the parameters of a finite autoregressive process. *Ann. Statist., 10.* 442-453.

Comer, John P., Jr. (1964). Some stochastic approximation procedures for use in process control. *Ann. Math. Statist., 35* 1137-1146.

Derman, C. and Sacks, J. (1959). On Dvoretzky's stochastic approximation theorem. *Ann. Math. Statist.*, *30* 601-605.

Fabian, V. (1978). On asymptotically efficient recursive estimation. *Ann. Statist.*, *6* 854-867.

Holst, Ulla (1980). Convergence of a recursive stochastic algorithm with m-dependent observations. *Scand. J. Statist.* *7* 207-215.

Holst, Ulla (1982). Convergence of a recursive stochastic algorithm with strongly regular observations. Technical Report. Department of Mathematical Statistics. University of Lund and Lund Institute of Technology.

Kushner, H.J. and Clark, D.S. (1978). *Stochastic Approximation Methods for Constrained and Unconstrained Systems*. Springer-Verlag. New York. 260 p.

Martin, R.D. and Masreliez, C.J. (1975). Robust estimation via stochastic approximation. *IEEE Trans. Inform. Theory IT-21* 263-271.

McLeish, D.L. (1975). A maximal inequality and dependent strong laws. *Ann. Prob.* *3* 829-839.

Parthasarathy, K.R. (1967). *Probability Measures on Metric Spaces*. Academic Press, New York.

Pham, Ruan D. and Tran, Lanh T. (1980). The strong mixing property of the autoregressive moving average time series model. Technical Report, Department of Mathematics, Indiana University.

Ranga, R.R. (1962). Relations between weak and uniform convergence of measures with applications. *Ann. Math. Statist.* *33* 659-680.

Robbins, H. and Monro, S. (1951). A stochastic approximation method. *Ann. Math. Statist.* *22* 400-407.

Robbins, H. and Siegmund, D. (1971). A Convergence Theorem for Nonnegative Almost Supermartingales and Some Applications. In *Optimizing Methods in Statistics* (J.S. Rustagi, ed.) 233-257. Academic Press, New York.

Ruppert, D. (1979). A new dynamic stochastic approximation procedure. *Ann. Statist.* *7* 1179-1195.

Ruppert, D. (1981). Stochastic approximation of an implicitly defined function. *Ann. Statist.* *9* 555-566.

Sequential probability ratio tests for homogeneous Markov chains

N. Schmitz and B. Süselbeck

Institut für Mathematische Statistik
Westfälische Wilhelms-Universität Münster
D-4400 Münster, Einsteinstraße 62

1. Introduction; simple hypotheses

Already A. Wald mentioned (see [10] p. 130) that his sequential probability ratio test (SPRT) may be defined also in the case of two simple hypotheses

$$H:\ P^X = P_i;\ i=1,2,$$

on a (time-discrete[1]) stochastic process

$$(\Omega, \mathcal{S}, P;\ X = (X_n)_{n\in\mathbb{N}_0})$$

in just the same way as in the i.i.d. case - merely the representability of the likelihood-ratio

$$(1)\qquad q(x_0,\dots,x_n) := \frac{f_2(x_0,\dots,x_n)}{f_1(x_0,\dots,x_n)},\ f_i \text{ densities,}$$

as a product of the single likelihood-ratios is lost:

For suitable constants $0<k_1<1<k_2<\infty$ an SPRT $\delta_{k_1,k_2} = (N_{k_1,k_2}, \varphi)$ is defined by the stopping rule

$$(2)\qquad N_{k_1,k_2} := \inf\{n\in\mathbb{N}_0:\ q(x_0,\dots,x_n) \notin (k_1;k_2)\}$$

($\inf\emptyset = \infty$) and the decision rule $\varphi = (\varphi_n)_{n\in\mathbb{N}_0}$ where

$$(3)\qquad \varphi_n(x_0,\dots,x_n) \cong \begin{cases} d_1 \\ d_2 \end{cases} \text{if } q(x_0,\dots,x_n) \begin{matrix} < \\ > \end{matrix} 1$$

and d_i denotes the decision in favour of H_i, i=1,2.

But simple examples show that even the termination property is lost in general:

(1.1) Example

Let two simple hypotheses on a homogeneous Markov chain with two states be given by the initial probabilities

1) For stochastic processes with general time-space some measurability conditions have to be fulfilled.

$$^1p = \binom{0}{1} = {^2p}$$

and the transition matrices

$$^1\underline{P} = \begin{pmatrix} 1 & 0 \\ 0.4 & 0.6 \end{pmatrix}, \quad ^2\underline{P} = \begin{pmatrix} 1 & 0 \\ 0.3 & 0.7 \end{pmatrix}.$$

These hypotheses coincide for the absorbing state 1; therefore every SPRT δ_{k_1,k_2} with

$$0 < k_1 < 3/4, \quad 1 < k_2 < \infty$$

is not closed, but $P_i(N_{k_1,k_2} = \infty) > 0$ for i=1,2.

Therefore also the exponential boundedness and the relations between the constants k_i and the error-probabilities

$$\alpha_i(\delta_{k_1,k_2}) := P_i(\varphi_{N_{k_1,k_2}} \cong d_{3-i}), \quad i=1,2$$

are lost in general.

On the other hand it is possible to characterize those testing problems, for which all SPRT's are closed, in a rather satisfactory way; in particular it turns out that the orthogonality of P_1, P_2 is a necessary and sufficient condition for this termination property (see [7], [8]).

For the special case of a homogeneous Markov chain with a finite state space it can be shown that the condition

<u>(1.2)</u> For at least one[2)] H_i there exists for any achievable irreducible subchain a pair (j_o, ℓ_o) of states such that

$$^1p_{j_o,\ell_o} \neq {^2p_{j_o,\ell_o}}$$

is necessary and sufficient for both the termination property and the exponential boundedness of all SPRT's (see [6]).

Moreover, it turns out that the SPRT's are - despite the fact that there does not exist in general any test which is uniformly optimal with respect to the expected sample size - of practical interest for homogeneous Markov chains with finite state space:

<u>(1.3)</u> Remark ([6])

Under the assumption (1.2) it may be expected that the SPRT's save about 70% of the sample size of corresponding fixed sample size tests.

---

2) Then the condition is fulfilled for the other hypothesis, too.

On the other hand, from a practical point of view the problem of testing two simple hypotheses seems to be rather artifical - instead one is usually faced with composite hypotheses

$$H_i: P^X \in \mathcal{P}_i, \; i=1,2$$

where $\mathcal{P}_i = \{P_\theta: \theta \in \Theta_i\}$ are disjoint families of distributions.

It is possible to define SPRT's also for this case by selecting two special distributions $P_i \in \mathcal{P}_i$ and using these for the construction (1) - (3). On the other hand it seems not to be obvious whether this yields reasonable tests.

2. Two examples with composite hypotheses

To show what may happen even with the termination property we consider two simple examples which partially go back to Strehl [9].

(2.1) Example

Let $X_0, X_1, \ldots$ be observations of a system whose possible states are 1,2 and 3. Let H_1, H_2 be the simple hypotheses that the X_n are iid with

$$P_1(X_n=j) = \begin{cases} 3/9 & j=1 \\ 4/9 \quad \text{for} & j=2; \\ 2/9 & j=3 \end{cases} \quad P_2(X_n=j) = \begin{cases} 2/9 & j=1 \\ 4/9 \quad \text{for} & j=2 \\ 3/9 & j=3 \end{cases}.$$

Then every SPRT is closed (and exponentially bounded) for all iid X_n with $P^{X_n} \neq \delta_2$.

But let in fact the X_n form a homogeneous Markov chain characterized by the initial distribution

$$p = (1/4, 1/2, 1/4)^T$$

and the transition matrix

$$\underline{P} = \begin{pmatrix} 0 & 0 & 1 \\ 1/3 & 2/3 & 0 \\ 1/3 & 2/3 & 0 \end{pmatrix}$$

(p is just the stationary distribution associated with $\underline{P}$). Then every SPRT δ_{k_1,k_2} with

$$k_1 < 2/3, \; k_2 > 3/2$$

will never terminate (with probability 1), i.e. the termination property is destroyed by the Markov dependence.

(2.2) Example

Let two distributions concerning a homogeneous Markov chain $X_o, X_1, \dots$ with 3 states be given by the initial distributions

$$^1p = (^1/_4, ^1/_2, ^1/_4)^T = {}^2p$$

and the transition matrices

$$^1\underline{P} = \begin{pmatrix} \frac{\varepsilon_1}{2} & \frac{\varepsilon_1}{2} & 1-\varepsilon_1 \\ \frac{1-\varepsilon_2}{3} & \frac{2}{3} & \frac{\varepsilon_2}{3} \\ \frac{1-\varepsilon_2}{3} & \frac{2}{3} & \frac{\varepsilon_2}{3} \end{pmatrix} \qquad ^2\underline{P} = \begin{pmatrix} \frac{\varepsilon_2}{2} & \frac{\varepsilon_2}{2} & 1-\varepsilon_2 \\ \frac{1}{3} & \frac{2}{3} & 0 \\ \frac{1-\varepsilon_1}{3} & \frac{2(1-\varepsilon_1)}{3} & \varepsilon_1 \end{pmatrix}$$

where $0<\varepsilon_1<\varepsilon_2<1$. These $^i\underline{P}$ are irreducible and aperiodic with $^1p_{j\ell} \neq {}^2p_{j\ell}$ $\forall (j,\ell) \neq (2,2)$; the corresponding distributions P_i are orthogonal. But let now the "true" distribution be the same as in (2.1). That $\underline{P}$ is irreducible and aperiodic, too; P is orthogonal to both P_i. The conditional distribution of

$$Z_n := \ell n\, ({}^2p_{X_{n-1},X_n} \; / \; {}^1p_{X_{n-1},X_n}),\; n \in \mathbb{N},$$

given $X_{n-1} = i$, is non-degenerate for i=2 and 3. Nevertheless every SPRT δ_{k_1,k_2} with

$$k_1 < 1 - \varepsilon_2, \quad k_2 > {}^1/_{(1-\varepsilon_2)}$$

will (with probability 1) never terminate.

Therefore the termination property (and the exponential boundedness) of the SPRT's is not "robust" against deviations from the hypothetical distributions P_i. □

(2.3) Remarks

a) The examples (2.1) and (2.2) turn out to be counterexamples against propositions of Phatarfod ([5], lemmata 2,3 and 4).

b) Condition (V.2) of Küchler [2] really avoids the difficulties arising in these examples, but it seems to be hard to verify that condition without knowing the true distribution.

3. On the termination property of SPRT's for homogeneous Markov chains

Using a degeneracy concept for matrices due to Miller [3] it is on the other hand possible to rectify and complete the results of Phatarfod:

(3.1) Lemma

Let $X = (X_n)_{n \in \mathbb{N}_0}$ be an irreducible homogeneous Markov chain with state space $M = \{1,\dots,m\}$, let $\underline{P}$, ${}^1\underline{P}$, ${}^2\underline{P}$ be hypothetical transition matrices[3)] of X and $P_{(i)}$ the corresponding distributions. Then every SPRT for P_1 against P_2 is closed under P if and only if for

$$\underline{P}(t) := (p_{j\ell} e^{t h_{j\ell}}) \text{ where } h_{j\ell} := \ln \frac{{}^2p_{j\ell}}{{}^1p_{j\ell}}, \ t \in \mathbb{C},$$

holds

(4) $\left\{ \begin{array}{ll} \text{(i)} & \underline{P}(t) \text{ is non-degenerate}^{4)} \\ \text{or} & \\ \text{(ii)} & \underline{P}(t) \text{ is degenerate with } c \neq 0. \end{array} \right.$

Proof: a) Let $\underline{P}(t)$ be degenerate with c = 0 and let

$$k := \max_{j,\ell \in M} e^{c_j - c_\ell}.$$

Then every SPRT δ_{k_1,k_2} with $k_1 < \frac{1}{k}$, $k < k_2$ will never terminate (with probability 1).

b) Let $\underline{P}(t)$ be degenerate with $c \neq 0$. Then

$$\prod_{j=1}^{n} \frac{{}^2p_{X_{j-1},X_j}}{{}^1p_{X_{j-1},X_j}} = e^{nc}\, e^{c_{X_0} - c_{X_n}} \qquad P\text{-a.s.} \quad \forall n \in \mathbb{N};$$

this yields the closedness of every SPRT.

c) Let P(t) be non-degenerate and

$$E_P(e^{tZ_1} \mid X_0 = j) < \infty \quad \forall t \in \mathbb{R},\ j \in M,$$

where

$$Z_n := \ln({}^2p_{X_{n-1},X_n} \,/\, {}^1p_{X_{n-1},X_n}), \quad n \in \mathbb{N}.$$

Then the proof of Küchler [2] yields the termination property.

d) Let there exist a $j_0 \in M$ (and a $t_0 \neq 0$) such that

$$\exp(t_0 \ln {}^2p_{j_0,X_1} \,/\, {}^1p_{j_0,X_1})$$

is not integrable; i.e. there exists, without loss of generality, an $\ell_0 \in M$ such that

---

3) To avoid trivial exceptions the initial probabilities are assumed to coincide: $p = {}^1p = {}^2p$.

4) $\underline{P}(t)$ is called degenerate if there exist $c, c_j \in \mathbb{R}$ such that $p_{j\ell}(t) = p_{j\ell} e^{t(c + c_j - c_\ell)}$ for alle j,ℓ,t.

$$p_{j_0,\ell_0} > 0 \text{ and } {}^2p_{j_0,\ell_0} > 0 = {}^1p_{j_0,\ell_0} ,$$

and moreover an $n_0 \in \mathbb{N}$ such that

$$\gamma := \min_{j \in M} \max_{1 < n < n_0} p^{(n)}_{j,j_0} \, p_{j_0,\ell_0} > 0.$$

This yields for every SPRT δ_{k_1,k_2}

$$P(N_{k_1,k_2} > kn_0) < (1-\gamma)^k \quad \forall k \in \mathbb{N}$$

and therefore the closedness of δ_{k_1,k_2}. □

In the cases b), c), d) one obtains

$$\lim_{n \to \infty} n^r \, P(N_{k_1,k_2} > n) = 0 \quad \forall r \in \mathbb{N}$$

and therefore

<u>(3.2)</u> <u>Corollary</u>

Under the assumptions of lemma (3.1) every SPRT is exponentially bounded for P if and only if condition (4) is fulfilled.

But it seems not be easy to compute, for given ${}^i\underline{P}$, those irreducible $\underline{P}$ for which condition (4) is violated. From the statistical point of view therefore the problem arises to develope sufficient conditions for (4) which do not need the (exact) knowledge of $\underline{P}$ - a very simple condition of this kind is that the elements of ${}^i\underline{P}$ and $\underline{P}$ are a priori known to be all positive.

Lemma (3.1) is concerned only with finite state spaces. But that result can partially be carried over to a special class of Markov process with infinitely many states: Let $X = (X_n)_{n \in \mathbb{N}_0}$ be an irreducible homogeneous Markov chain with state space $M = \mathbb{N}$ and let $\underline{P}$, ${}^1\underline{P}$, ${}^2\underline{P}$ be hypothetical transition matrices. We will assume

(5) X is positive recurrent (under $\underline{P}$)[5)]

If the sequence

5) For irreducible homogeneous Markov chains with finite state space this assumption is automatically fulfilled.

$$(E(e^{Z_1 t_o} | X_o = j))_{j \in M}$$

is unbounded (for suitable $t_o \in \mathbb{R}$), then there exists, for given constants k_1, k_2, a state j_o such that

$$P(|Z_1| > \ln(k_2/k_1) | X_o = j_o) > 0.$$

The positive recurrence yields therefore the termination property of the SPRT's δ_{k_1,k_2}.

We may therefore restrict attention to the case

(6) $$E(e^{Z_1 t} | X_o = j) < M_t < \infty \qquad \forall j \in M, t \in \mathbb{R}.$$

Then the matrix $\underline{P}(t)$ defines a bounded linear operator on ℓ^∞; for all $t \in \mathbb{C}$ and $n \in \mathbb{N}$ one obtains for the (j,ℓ)-th element of $(\underline{P}(t))^n$:

$$(P(t)^n)_{j,\ell} = p_{j,\ell}^{(n)} \, E(e^{t S_n} | X_o = j, X_n = \ell)$$

where

$$S_n := \sum_{j=1}^{n} Z_j.$$

Moreover, for every $n \in \mathbb{N}$ the function

$$\underline{P}^n(t) := (\underline{P}(t))^n$$

is holomorphic on $\mathbb{C}$ and its k-th derivative at $t \in \mathbb{C}$ is

(7) $$((P^n(t))^{(k)})_{j,\ell} = p_{j,\ell}^{(n)} \, E(S_n^k \, e^{t S_n} | X_o = j, X_n = \ell);$$

the matrices $(P^n(t))^{(k)}$ define bounded linear operators on ℓ^∞, too.

Since $p_{j,\ell}(t) > 0$ if and only if $p_{j,\ell}(0) = p_{j,\ell} > 0$, the matrix $\underline{P}(t)$, $t \in \mathbb{R}$, is irreducible and has the same period d as $\underline{P}$.

The spectral radius $\mathrm{spr}(\underline{P}) = 1 =: \lambda_1$ of $\underline{P}$ is an eigenvalue as well of $\underline{P}$ as of $\underline{P}^T$; if the period of $\underline{P}$ is d then

$$\lambda_j := \exp(2\pi i \tfrac{j-1}{d}), \quad 1 < j < d,$$

are (simple) eigenvalues of $\underline{P}$. We will assume

(8) There exist a neighbourhood $G \subset \mathbb{C}$ of 0 and functions $\lambda: G \longrightarrow \mathbb{C}$, $r: G \longrightarrow \ell^{\infty}$ such that

(i) for all $t \in G$ holds: $\lambda(t) = \operatorname{spr} \underline{P}(t)$, $\lambda(t)$ is an eigenvalue of $\underline{P}(t)$ with right eigenvector $r(t)$, $r_j(t) > 0$ for all $j \in \mathbb{N}$, and $\lambda(0) = 1$, $r(0) = (1,1,\dots)^T$,

(ii) λ and r are holomorphic on G.

Assumption (8) is e.g. fulfilled (Oeljeklaus [4]) if

(9) $$\beta^* := \sup\{|\lambda| : \lambda \in \sigma(\underline{P}),\ \lambda \notin \{\lambda_1,\dots,\lambda_d\}\} < 1$$

where $\sigma(\underline{P})$ denotes the spectrum of $\underline{P}$ [6].

(10) $$\lambda_1'(0) = 0 \implies \lambda_1''(0) > 0$$ [7].

These assumptions are sufficient to extend the results of Küchler to the denumerable case:

<u>(3.3)</u> <u>Theorem</u>

Let $X = (X_n)_{n \in \mathbb{N}_0}$ be an irreducible, homogeneous Markov chain with state space $\mathbb{N}$; let $\underline{P}$, ${}^1\underline{P}$, ${}^2\underline{P}$ be hypothetical transition matrices of X and $P_{(i)}$ the corresponding distributions (see footnote 3)); let X be positive recurrent and assume either

(a) $(E(e^{Z_1 t_0} | X_0 = j))_{j \in \mathbb{N}}$ is unbounded (for suitable t_0)

or

(b) conditions (6), (8), (10) are fulfilled

under $\underline{P}$. Then every SPRT for P_1 against P_2 is closed under P.

<u>(3.4)</u> <u>Lemma</u>

Under the assumptions of (3.3) for every SPRT δ_{k_1,k_2} holds $E(N_{k_1,k_2}) < \infty$.

---

6) (9) is always fulfilled for finite state spaces.

7) According to Miller [3] and Keilson/Wishart [1] this assumption is in the finite case fulfilled for non-degenerate matrices $\underline{P}$.

According to the footnotes 5), 6), 7) these propositions are natural extensions of the finite case.

Proof: In case (b) of (3.3) holds

$$(\lambda(t))^n r(t) = \underline{P}^n(t)r(t) \qquad \forall n \in \mathbb{N}$$

and thus

$$(\lambda(t))^n p\cdot r(t) = p\cdot P^n(t)r(t)$$

for the initial distribution $p = (p_1, p_2 \dots)$.
Denoting $\alpha(t) = p\cdot r(t)$ one obtains, according to (7), for the first derivatives of $p\cdot P^n(t)r(t)$:

(i) $$[p\cdot\underline{P}^n(t)r(t)]' = p\,[\underline{P}^n(t)]'r(t) + p\cdot\underline{P}^n(t)r'(t)$$
$$= E(S_n e^{tS_n} r_{X_n}(t)) + E(e^{tS_n} r'_{X_n}(t))$$

(ii) $$[p\cdot\underline{P}^n(t)r(t)]'' = p\cdot[\underline{P}^n(t)]''r(t)+2p\cdot[P^n(t)]'r'(t)+p\underline{P}^n(t)r''(t)$$
$$= E(S_n^2\, e^{tS_n} r_{X_n}(t))+2E(S_n e^{tS_n} r'_{X_n}(t))+E(e^{tS_n} r''_{X_n}(t)).$$

Since on the other hand

$$[\lambda(t)^n\alpha(t)]'' = n(\lambda(t))^{n-1}\lambda''(t)\alpha(t)+n(n-1)(\lambda(t))^{n-2}(\lambda'(t))^2\alpha(t)$$
$$+ 2n(\lambda(t))^{n-1}\lambda'(t)\alpha'(t)+(\lambda(t))^n\alpha''(t)$$

it follows from $\lambda(0) = 1$, $\alpha(0) = 1$, $r_j(0) = 1 \ \forall j \in \mathbb{N}$ and (ii) that

$$n\lambda''(0)+n(n-1)(\lambda'(0))^2+2n\lambda'(0)\alpha'(0)+\alpha''(0) = E(S_n^2)+2E(S_n r'_{X_n}(0))+E(r''_{X_n}(0)).$$

According to assumption (10) the left hand side of this equation is divergent. From

$$|r'_{X_n}(0)| < ||r'(0)|| < \infty, \ |r''_{X_n}(0)| < ||r''(0)|| < \infty .$$

$$E(S_n^2) > (E|S_n|)^2 \qquad \forall n \in \mathbb{N}$$

one therefore obtains

$$\lim_{n\to\infty} E(S_n^2) = \infty.$$

This yields together with the positive recurrence the propositions (3.3),(3.4). □

4. A "fundamental identity"

To get approximations for the OC-function of SPRT's Phatarfod [5] and Küchler [2] derived an analogon to Wald's "fundamental identity". The following martingale-theoretical proof seems to be simpler than their arguments and moreover yields a result also for a class of Markov chains with denumerable state space:

(4.1) Theorem

Let $X = (X_n)_{n \in \mathbb{N}_0}$ be an irreducible, homogeneous Markov chain, let $\underline{P}$, $^1\underline{P}$, $^2\underline{P}$ be hypothetical transition matrices of X and $P_{(i)}$ the corresponding distributions, and let δ_{k_1,k_2} be an SPRT. If the conditions (5), (6) and (8)(i) are fulfilled and δ_{k_1,k_2} is closed, then for all $t \in \mathbb{R}$ with $\lambda(t) > 1$ holds

$$E\left(\frac{1}{\lambda(t)^{N_{k_1,k_2}}} e^{tS_{N_{k_1,k_2}}} r_{X_{N_{k_1,k_2}}}(t)\right) = p \cdot r(t) .$$

The proof of (4.1) makes use of the following

(4.2) Lemma

Let $\hat{Y}_n(t)$ be defined by

$$Y_n(t)(\omega) := e^{tS_n(\omega)} r_{X_n(\omega)}(t)$$
$$\hat{Y}_n(t)(\omega) := \frac{1}{p \cdot r(t)(\lambda(t))^n} Y_n(t)(\omega) \qquad \omega \in \Omega,\ n \in \mathbb{N},$$

and denote

$$\mathcal{F}_n := \sigma(X_0, \ldots, X_n), \quad n \in \mathbb{N}_0.$$

Then $(\hat{Y}_n(t), \mathcal{F}_n)_{n \in \mathbb{N}}$ is a martingale and

$$E(\hat{Y}_n(t)) = 1 \quad n \in \mathbb{N}.$$

Proof of (4.2): For $n \in \mathbb{N}$ holds

$$\begin{aligned}
E(\hat{Y}_{n+1}(t) \mid \mathcal{F}_n) &= \frac{1}{p \cdot r(t)(\lambda(t))^{n+1}} E(e^{tS_{n+1}} \mid \mathcal{F}_n) \\
&= \frac{1}{p \cdot r(t)(\lambda(t))^{n+1}} e^{tS_n} E(e^{tZ_{n+1}} \mid \mathcal{F}_n) \\
&= \frac{1}{p \cdot r(t)(\lambda(t))^{n+1}} e^{tS_n} E(e^{tZ_{n+1}} \mid X_n) \\
&= \frac{1}{p \cdot r(t)(\lambda(t))^{n+1}} e^{tS_n} \sum_{\ell \in M} e^{th_{X_n,\ell}} p_{X_n,\ell} \\
&= \frac{1}{p \cdot r(t)(\lambda(t))^{n+1}} e^{tS_n} \lambda(t) r_{X_n}(t) = \hat{Y}_n(t).
\end{aligned}$$

From

$$E(Y_1(t)) = p\cdot\underline{P}(t)r(t) = \lambda(t)p\cdot r(t)$$

follows $E(\hat{Y}_1(t)) = 1$ and therefore the remaining part of (4.2). □

Proof of (4.1): N_{k_1,k_2} is by assumption a stopping rule with respect to $(\mathcal{Y}_n)_{n\in\mathbb{N}_0}$ with $N_{k_1,k_2} > 1$. For $\lambda(t) > 1$ holds

$$\int_{\{N_{k_1,k_2} > n\}} Y_n(t)dP <$$

$$< \begin{cases} \dfrac{1}{p\cdot r(t)(\lambda(t))^n} k_2^t \|r(t)\| P(\{N_{k_1,k_2} > n\}) & \text{for} \quad t > 0 \\[2ex] \dfrac{1}{p\cdot r(t)(\lambda(t))^n} k_1^t \|r(t)\| P(\{N_{k_1,k_2} > n\}) & \text{for} \quad t < 0 \end{cases}$$

$$\xrightarrow[n\to\infty]{} 0.$$

The optional sampling theorem therefore yields

$$E(\hat{Y}_{N_{k_1,k_2}}(t)) = 1$$

and therefore the assertion of theorem (4.1). □

Under some assumptions on $\lambda(t)$, which are in the finite case fulfilled under condition (V.2) of Küchler [2], this fundamental identity may be used to derive the OC-function and the ASN-function of SPRT's and to give, neglecting the excesses over the boundaries, approximations of these functions.

References

[1] Keilson, J., Wishart, D. (1964): A central limit theorem for processes defined on a finite Markov chain.
Proc. Cambridge Phil. Soc. 60; 547-567.

[2] Küchler, I. (1978): Der sequentielle Quotiententest bei irreduziblen homogenen Markovschen Ketten mit endlichem Zustandsraum.
Math. Operationsforsch. Statist. 9; 227-239.

[3] Miller, H. (1961): A convexity property in the theory of random variables defined on a finite Markov chain.
Ann. Math. Statist. 32; 126o-127o.

[4] Oeljeklaus, G. (1981): Der Likelihood-Quotienten-Sequenztest bei zusammengesetzten Hypothesen über homogene Markov-Ketten.
Unpublished Diplomarbeit Münster.

[5] Phatarfod, R. (1965): Sequential analysis of dependent observations I.
Biometrika 52; 157-165.

[6] Schmitz, N. (1968): Likelihoodquotienten-Sequenztests bei homogenen Markoffschen Ketten.
Biometrische Z. 10; 231-247.

[7] - (1970): Existenz von sequentiellen Tests zu vorgegebenen Niveaus.
Archiv Math. XXI; 617-628.

[8] - (1973/77): Sequentielle Mehrentscheidungsverfahren mit vorgeschriebenem Irrtumsvektor.
Oper. Res. Verf. 17; 317-340; 24, 176-177.

[9] Strehl, R. (1978): Bemerkungen zur exponentiellen Beschränktheit des Likelihoodquotienten-Sequenztests bei Vorliegen von Abhängigkeiten.
Unpublished Diplomarbeit Münster.

[10] Wald, A. (1945): Sequential tests of statistical hypotheses.
Ann. Math. Statist. 16; 117-186.

ALLOCATION RULES FOR SEQUENTIAL CLINICAL TRIALS

D. Siegmund
Stanford University
Stanford, CA 94305/USA

Consider the following simplified model of a clincial trial. Patients arrive sequentially at a treatment center and receive one of two treatments: A or B. The (immediate response of the ith patient to receive treatment A is x_i, $i = 1,2,\ldots$, that of the jth patient to receive treatment B is y_j, $j = 1,2,\ldots$. At any stage of the process, having observed $x_1,\ldots,x_m$, $y_1,\ldots,y_n$, the experimenter can stop the experiment and declare (1) A is the better treatment, (2) B is better, or (3) there is essentially no difference between A and B; or he can continue the experiment and assign the next patient to treatment A or B according to some allocation rule. In this paper we shall be primarily interested in the experimenter's allocation rule, which should be selected insofar as possible (i) to permit valid inferences upon termination of the experiment and (ii) to minimize in some sense the number of patients receiving the inferior treatment during the course of the experiment.

The specific mathematical framework developed below to discuss this problem was introduced by Flehinger, Louis, Robbins, and Singer (1972) and developed by Robbins and Siegmund (1973), Louis (1975), and Hayre (1979). To a considerable extent the present paper is a review and exposition of these ideas. An interesting and somewhat different approach has been recently developed by Bather (1980, 1981), and it would be interesting to make a systematic comparison of Bather's approach with that outlined below.

We assume that $x_1,\ldots,x_m,\ldots$ are independent $\mathfrak{N}(\mu_1,1)$ and $y_1,\ldots,y_n,\ldots$ are independent $\mathfrak{N}(\mu_2,1)$ random variables, and that the x's and y's are independent. For an indication how the results given here can be extended via large sample approximations to non-normal data, see Robbins (1974). Let $\delta = \mu_1 - \mu_2$, and to be specific assume that the better treatment is that yielding the larger mean response. Hence to say that treatment A is superior to say that $\delta > 0$, etc.

Let $\bar{x}_m = m^{-1} \Sigma_{i=1}^m x_i$, $\bar{y}_n = n^{-1} \Sigma_{j=1}^n y_j$, and $z_{m,n} = \frac{mn}{m+n} (\bar{x}_m - \bar{y}_n)$. Having observed $x_1,\ldots,x_m$ and $y_1,\ldots,y_n$, the natural estimator of δ is $\hat{\delta}_{m,n} = \bar{x}_m - \bar{y}_n$. Since problems of statistical inference about δ are invariant under changes in location of the data, it is reasonable to consider invariant procedures, i.e. those based on the process $z_{m,n}$, $m,n \geq 1$ or equivalently on $u_i = x_i - x_1$, $i = 1,2,\ldots$ and $v_j = y_j - x_1$, $j = 1,2,\ldots$. Our first result is a "separation" theorem, which says that in a certain sense the problem of statistical inference about δ can be separated from the problem of allocation, provided we restrict ourselves to invariant procedures (cf. Lemma 1 of Robbins and Siegmund, 1973).

It is convenient to introduce the following notation:

$$\mathfrak{F}_{m,n} = \mathfrak{B}(u_i,\ i \leq m;\ v_j,\ j \leq n) ,$$

$W(t)$ = Brownian motion with drift δ ,

$\mathcal{F}_W(t) = \mathcal{B}(W(s),\ s \le t)$,

$t_{m,n} = mn/(m+n)$.

Proposition 1. For arbitrary $m,n \ge 1$

$$\mathcal{L}(z_{m+1,n} - z_{m,n} | \mathcal{F}_{m,n}) = \mathcal{L}(W(t_{m+1,n}) - W(t_{m,n}) | \mathcal{F}_W(t_{m,n}))$$

and

$$\mathcal{L}(z_{m,n+1} - z_{m,n} | \mathcal{F}_{m,n}) = \mathcal{L}(W(t_{m,n+1}) - W(t_{m,n}) | \mathcal{F}_W(t_{m,n})) .$$

Proof. Simple algebra yields

$$z_{m+1,n} - z_{m,n} = nx_{m+1}/(m+n+1) - n(\sum_1^n y_j + \sum_1^m x_i)/(m+n)(m+n+1) . \tag{1}$$

It is easy to see that the two terms on the right hand side of (1) are each uncorrelated with u_i, $i \le m$ and v_j, $j \le n$. Hence by properties of the normal distribution $z_{m+1,n} - z_{m,n}$ given $\mathcal{F}_{m,n}$ is normally distributed with expectation

$$\begin{aligned} & n\mu_1/(m+n+1) - n(m\mu_1 + n\mu_2)/(m+n)(m+n+1) \\ &= n^2\delta/(m+n)(m+n+1) = \delta(t_{m+1,n} - t_{m,n}) \end{aligned}$$

and variance

$$[n/(m+n+1)]^2 + n^2/(m+n)(m+n+1)^2 = t_{m+1,n} - t_{m,n} .$$

But this is precisely the conditional distribution of $W(t_{m+1,n}) - W(t_{m,n})$ given $\mathcal{F}_W(t_{m,n})$. A similar argument applies to $z_{m,n+1} - z_{m,n}$.

Corollary 1. For any invariant allocation rule the processes

$$\{z_{m,n} - mn\ /(m+n),\ \mathcal{F}_{m,n}\}$$

and

$$\{(z_{m,n} - mn\ /(m+n))^2 - mn/(m+n),\ \mathcal{F}_{m,n}\}$$

are martingales.

An important consequence of Proposition 1 is that for any allocation rule based on the process $z_{m,n}$ or equivalently on the u's and v's, there exists an "isomorphic" allocation rule based on $W(t_{m,n})$ such that the sequences of pairs

$\{z_{m,n}, mn/(m+n)\}$ and $\{W(t_{m,n}), t_{m,n}\}$ have the same joint distribution. Hence if $0 < t_f \le \infty$ various allocation rules yield sequences of "observations" $W(t_{m,n})$, $0 < t_{m,n} \le t_f$, which differ only in the (random) times at which the Brownian path is observed. Because the Brownian paths are continuous and the increments $t_{m+1,n} - t_{m,n}$ or $t_{m,n+1} - t_{m,n} \in (0,1)$, the exact choice of allocation rule has a limited effect on the joint distribution of the observed data provided $t_{m,n} \to t_f$ as $m+n \to \infty$. In particular, if the process $\{W(t_{m,n}), t_{m,n}\}$ is observed until it first leaves some region with a continuous boundary in the space time plane, the point at which the process leaves the region has a distribution which is approximately independent of the allocation rule.

For our present purposes this has the following consequences. Assume temporarily that the allocation rule does not depend on the data -- for example that observations are taken in pairs (x_i, y_i), $i = 1,2,\ldots$ and $w_i = x_i - y_i$. Assume also that in this context we favor a particular procedure for making inferences about $\delta = Ew_i$. For example, to test H_0: $\delta = 0$ we stop sampling at $\min(T, 2\nu)$, where

$$T = \inf\{n: |\sum_1^n w_i| \ge 2b\}$$

and ν is some positive integer. If $T \le 2\nu$ and $\Sigma_1^T m_i \ge 2b$ we reject H_0 and say that $\delta > 0$; if $T \le 2\nu$ and $\Sigma_1^T w_i \le -2b$ we reject H_0 and say that $\delta < 0$; if $T > 2\nu$ we accept H_0 as being approximately true. The power function of this test of H_0: $\delta = 0$ against H_1: $\delta \neq 0$ is $P_\delta\{T \le 2\nu\}$ and the expected sample size is $E_\delta(T \wedge 2\nu)$. For any invariant allocation rule there exists the analogous procedure: stop sampling when $mn/(m+n) \ge \nu$ or at

$$(\tilde{M}, \tilde{N}) = \inf\{(m,n): |z_{m,n}| \ge b\} \quad ,$$

whichever occurs first, and reject H_0: $\delta = 0$ if and only if $\tilde{M}\tilde{N}/(\tilde{M} + \tilde{N}) \le \nu$. It follows from the remarks following Proposition 1 that the power function $P_\delta\{\tilde{M}\tilde{N}/(\tilde{M} + \tilde{N}) \le \nu\}$ of this test satisfies

$$P_\delta\{\tilde{M}\tilde{N}/(\tilde{M} + \tilde{N}) \le \nu\} \cong P_\delta\{T \le 2\nu\} \quad , \tag{2}$$

and

$$2E_\delta\{[\tilde{M}\tilde{N}/(\tilde{M} + \tilde{N})] \wedge \nu\} \cong E_\delta(T \wedge 2\nu) \quad . \tag{3}$$

Hence we have obtained a sequential test whose power function is to a considerable extent independent of the allocation rule used, and we are free to consider different allocation rules in an attempt to minimize the number of observations taken on the inferior treatment.

Note that although we proceed with the discussion for one particular sequential test, we could equally well consider others, e.g. a repeated significance test.

Before we consider in detail the choice of allocation rule, it is helpful to observe the following limits imposed by (3). Let M and N denote the number of x and y observations respectively when sampling stops, so $MN/(M+N) \cong [\tilde{M}\tilde{N}/(\tilde{M}+\tilde{N})] \wedge \nu$. Since $\min(M,N) \geq MN/(M+N)$ with equality if and only if $\max(M,N) = \infty$, it follows from (3) that

$$\min(E_\delta M, E_\delta N) \geq \frac{1}{2} E_\delta(T \wedge 2\nu) \quad , \tag{4}$$

and a necessary condition for approximate equality in (4) is that $\max(E_\delta M, E_\delta N)$ be extremely large. Since $x(1-x) \leq 1/4$ with equality if and only if $x = 1/2$, by (3)

$$E_\delta(M+N) \geq 4E_\delta\{MN/(M+N)\} \cong 2E_\delta(T \wedge 2\nu) \tag{5}$$

and there is equality in (5) if and only if M and N are approximately equal with probability one. From (4) and (5) we conclude that the expected number of observations on the inferior treatment is at least $1/2$ as large as pairwise allocation requires; and any deviation from pairwise allocation results in some increase in the total expected sample size.

The following argument for choosing an allocation rule is due to Hayre (1979). Suppose that when δ is the true difference in mean response, the cost to the experimenter of an x observation is $g(\delta)$ while that of a y observation is $h(\delta)$. Hence the total expected cost of sampling is

$$g(\delta)E_\delta(M) + h(\delta)E_\delta(N) \quad . \tag{6}$$

The overall risk function is the sum of (6) and the risk associated with making a wrong terminal decision. But since the power function of our test is essentially independent of the allocation rule used, we can ignore the terminal decision part of the risk function and attempt to minimize (6). Since

$$M = [MN/(M+N)](1+M/N) = [MN/(M+N)](1+Q) \quad , \tag{7}$$

say, and

$$N = [MN/(M+N)](1+Q^{-1}) \quad , \tag{8}$$

we can rewrite (6) as

$$[h+g]E_\delta[MN/(M+N)] + E_\delta\{[MN/(M+N)][gQ+hQ^{-1}]\} \quad . \tag{9}$$

Moreover, by (3) the first term in (9) is essentially independent of the allocation rule, so we attempt to minimize the second. Calculus shows that for every Q $gQ + hQ^{-1} \geq 2(gh)^{1/2}$ with equality if and only if

$$Q = (h/g)^{1/2} \quad . \tag{10}$$

Hence a lower bound to (9) is

$$(h^{1/2} + g^{1/2})^2 \; E_\delta[MN/(M+N)] \quad , \tag{11}$$

which could be achieved only if we could allocate observations so that

$$P_\delta\{N/M = [g(\delta)/h(\delta)]^{1/2}\} = 1 \quad . \tag{12}$$

Since δ is unknown this is impossible, but as an approximation we consider the allocation rule which takes the next observation from the y population if and only if

$$n/m < [g(\bar{x}_m - \bar{y}_n)/h(\bar{x}_m - \bar{y}_n)]^{1/2} \quad . \tag{13}$$

To the extent that this allocation rule behaves as we hope it will, i.e. to the extent that (12) is approximately true, by (7) and (8) we have the approximations

$$E_\delta(M) \cong E_\delta[MN/(M+N)][1 + (h/g)^{1/2}] \quad , \tag{14}$$

$$E_\delta(N) \cong E_\delta[MN/(M+N)][1 + (g/h)^{1/2}] \quad , \tag{15}$$

and the risk (6) is approximately the lower bound (11).

A numerical example illustrating these results is given in Table 1. The functions g and h are of the form

$$g(\delta) = h(-\delta) = \begin{cases} 1 & \text{if} \quad \delta > 0 \\ 1 + d|\delta| & \text{if} \quad \delta < 0 \quad . \end{cases} \tag{16}$$

This choice has the interpretation that the basic (experimental) cost of an observation is unity, and the additional (ethical) cost of assigning the inferior treatment is proportional to $|\delta|$. For comparison the first row in each cell of Table 1 is for pairwise allocation, and the computations of power and expected sample size use the approximations suggested by Siegmund (1979) and shown to be very accurate. The second row of each cell gives results for the sampling rule (13) with g and h defined by (16). The first entry is the outcome of a 400 repetition Monte Carlo experiment, and the parenthetical entry is the theoretical approximation given by (15), (14), or (11). The Monte Carlo results lend support to our informal interpretation of Proposition 1 to the effect that the power and $E_\delta[MN/(M+N)]$ are approximately independent of the allocation rule, and they indicate that the approximations (15), (14), and (11) are quite good. Hayre (1979) reached the same conclusions for a different stopping rule and value of d in (16). Most importantly the results show a fairly substantial decrease in risk of about 15-30% when the allocation rule (13) is used.

TABLE 1

First row in each cell is for pairwise allocation; second row is for allocation rule (13) with h and g given by (16) with $d = 20$; in all cases $b = 10.8$, $\nu = 25$; theoretical calculations are in parentheses; others are Monte Carlo

δ	Power	$E_\delta(N)$	$E_\delta(M)$	$2E_\delta[MN/(M+N)]$	Risk
1.13	(1.00)	(19.9)	(19.9)	(19.9)	(490)
	1.00	12.3 (12.0)	53.7 (58.3)	20.0	343 (341)
.85	(.986)	(26.4)	(26.4)	(26.4)	(502)
	.995	15.9 (16.9)	62.7 (59.7)	25.2	349 (363)
.57	(.788)	(36.8)	(36.8)	(36.8)	(493)
	.793	24.0 (23.6)	77.5 (83.2)	36.3	375 (376)
.28	(.279)	(46.7)	(46.7)	(46.7)	(355)
	.298	36.4 (32.5)	77.4 (83.5)	46.6	318 (298)
.00	(.050)	(49.5)	(49.5)	(49.5)	
	.048	54.1	56.3	49.6	

The reduction in risk of 15-30% compared to pairwise sampling in Table 1 makes the allocation rule (13) seem attractive; but it is not large enough to overwhelm certain disadvantages without further investigation. (By way of comparison a fixed sample size with paired observations requires 48 pairs to achieve about the same power function as in Table 1. This leads to risks of 1181 and 912 for $\delta = 1.13$ and .85, so sequential sampling with pairwise allocation leads to a reduction in risk for large $|\delta|$ of about 50% compared to a fixed sample.) The disadvantages include (i) the fact that the allocation rule (13) is non-randomized, (ii) questionable performance when the patient population is stratified, (iii) difficulty in implementation if the data are examined occasionally, but not continuously, and (iv) questionable performance for survival data, where treatment assignments must be made for new patients before data become available on old ones. Accommodating these difficulties leads to some deterioration in the performance of adaptive allocation rules, which may lead to questioning their desirability at all. Here we discuss (i) and (ii) by means of a Monte Carlo experiment.

The advantages of randomization in clinical trials has been discussed at great length-primarily in an effort to eliminate selection bias (e.g. Blackwell and Hodges, 1957), but secondarily to provide the possibility of a permutation test of the hypothesis of no treatment effect. It is easy to define a randomized version of (13) to cut down on selection bias. For example, we might select treatment B or treatment A with relative probabilities given by the right hand side of (13). More precisely, let

$$\lambda_{m,n} = [g(\bar{x}_m - \bar{y}_n)/h(\bar{x}_m - \bar{y}_n)]^{1/2}$$

and take the next observation from the y population if and only if

$$U_{m+n+1} \leq \lambda_{m,n}/(1+\lambda_{m,n}) \quad , \tag{17}$$

where $U_1, U_2, \ldots$ is an auxiliary sequence of independent uniform random variables which we generate. Asymptotically this rule generates the appropriate relative frequencies of treatment selections. A more sophisticated version would be one which takes account of how far n/m is from the desired ratio of $\lambda_{m,n}$ in selecting the next treatment. For g and h given by (16) with $d = 20$, and for δ in the range $[-1,1]$, the right hand side of (17) is in the range $(.1,.9)$ with high probability, so there is always some indeterminacy in the next treatment assignment.

TABLE 2

Randomized Allocation (17)

$b = 10.8$, $\nu = 25$, g and h given by (16) with $d = 20$

δ	Power	$E_\delta(N)$	$E_\delta(M)$	2E [MN/(M+N)]	Risk
1.13	1.00	12.7	50.3	19.5	350
.85	.995	18.2	60.3	26.8	388
.57	.793	26.6	72.4	37.0	402
.28	.315	38.8	72.0	46.0	328

Table 2 gives the outcome of a 400 replication Monte Carlo experiment using the randomized allocation rule (17). By comparison with Table 1 we see that randomization has led consistently to an increase in risk, but one so slight that the benefits of randomization seem to outweigh the liability.

The question of stratification is more complicated because the results may depend on the number and relative sizes of different strata. The difficulties are most acute with a large number of small strata, where one usually wants to guarantee a certain amount of balance in the sample from each stratum, so that a stratum could be analyzed by itself if the model relating different strata seems to be inappropriate.

To be specific suppose there are r strata and for $k = 1,2,\ldots,r$, in stratum k the response of the ith patient on treatment A is x_{ki}, which is distributed $\mathcal{N}(\mu_k + \delta, 1)$, and that of the jth patient on treatment B is y_{kj}, distributed $\mathcal{N}(\mu_k, 1)$. After m_k assignments of treatment A and n_k of treatment B in the kth stratum, the maximum likelihood estimator of the treatment effect δ is (in the obvious notation)

$$\text{(18)} \qquad \hat{\delta}(\underline{m},\underline{n}) = \frac{\sum_{k=1}^{r} \frac{m_k n_k}{m_k + n_k} (\bar{x}_{k,m_k} - \bar{y}_{k,n_k})}{\sum_{k=1}^{r} \frac{m_k n_k}{m_k + n_k}}$$

Let $z(\underline{m},\underline{n})$ denote the numerator and $t(\underline{m},\underline{n})$ the denominator of $\hat{\delta}(\underline{m},\underline{n})$.

It is easy to obtain a result analogous to Proposition 1, and hence to conclude that $z(\underline{m},\underline{n})$ behaves like Brownian motion with drift δ in the time scale of $t(\underline{m},\underline{n})$ provided that an invariant treatment allocation rule is used. Here invariant means that the choice of the next treatment assignment may depend only on the vector of differences $(\bar{x}_{1,m_1} - \bar{y}_{1,n_1}, \ldots, \bar{x}_{r,m_r} - \bar{y}_{r,n_r})$. Hence as above we can test $\delta = 0$ with a test whose power function is essentially independent of the (invariant) allocation rule used, and we can turn our attention to the cost of sampling.

In analogy with (6) suppose the expected cost of sampling is given by

$$\text{(19)} \qquad g(\delta)\sum_k EM_k + h(\delta)\sum_k EN_k ,$$

where M_k (N_k) is the number of x's (y's) observed in the kth stratum, $k = 1,2,\ldots,r$. The argument leading to (11) now gives as a lower bound to (19)

$$\text{(20)} \qquad (h^{1/2} + g^{1/2})^2 E_\delta [\sum_{k=1}^{r} M_k N_k / (M_k + N_k)]$$

and there is equality between (19) and (20) if and only if (cf. (12))

$$\text{(21)} \qquad P_\delta\{N_k/M_k = [g(\delta)/h(\delta)]^{1/2}\} = 1 \quad \text{for all} \quad k = 1,\ldots,r \ .$$

This suggests in analogy with (13) that if a new patient arrives and falls into stratum k, then he is assigned treatment B if and only if

$$\text{(22)} \qquad n_k/m_k < [g(\hat{\delta}(\underline{m},\underline{n}))/h(\hat{\delta}(\underline{m},\underline{n}))]^{1/2}$$

where $\hat{\delta}$ is given in (18).

However, there is an additional practical consideration, which is especially important when there are small strata. The model of fixed treatment effect across strata is somewhat tentative and usually must be checked. To do this requires a minimal amount of balance in the assignment of treatments in each stratum individually. Hence we modify the sampling rule (22) by choosing some small positive number ν_0, and use (22) only if $m_k n_k/(m_k + n_k) \geq \nu_0$. If $m_k n_k/(m_k + n_k) < \nu_0$, we make the treatment assignment in some way that provides for about half of the first $4\nu_0$ patients to receive one treatment and half the other. Of course, this effects our ability to approximate (21), especially in small strata where the threshold ν_0 may not be exceeded; but it avoids the disastrous situation where almost all assignments in a small stratum are to one treatment.

Table 3 reports the results of a Monte Carlo experiment to determine the effects of stratification together with randomization. The test is defined by the same parameters as those in Tables 1 and 2, and hence has essentially the same power function. There are four strata in the relative sizes 4:3:2:1. The threshold is $\nu_0 = 3$, and strict pairwise sampling is used in each stratum until this threshold is reached. Thereafter a randomized version of (21) as specified in (17) is used.

The results are more ambiguous than in Table 2. For large $|\delta|$ stratification substantially increases the risk to the extent that sequential data dependent allocation seems only slightly better than pairwise sampling. For small $|\delta|$ there is a comparatively insignificant increase in risk. Although these results are not surprising qualitatively, and therefore probably persist to some extent under different

TABLE 3

Stratified Data, Randomized Allocation

$b = 10.8,\ \nu_0 = 3,\ \ = 25,\ d = 20$

δ	Power	$\sum_k E_\delta(N_k)$	$\sum_k E_\delta(M_k)$	$2\sum_k E_\delta[M_k N_k/(M_k+N_k)]$	Risk
1.13	1.00	17.6	23.8	19.8	439
.85	.990	22.3	35.1	26.6	449
.57	.848	29.2	47.8	35.2	409
.28	.288	40.3	61.1	46.9	327

experimental conditions, the exact magnitude of the changes may well be sensitive to the number and size of the strata, the parameters b, ν_0, ν, etc.

A conclusion to be drawn from Tables 2 and 3 is that practical constraints on using an allocation rule like (13) may reduce the advantage over pairwise allocation exhibited in Table 1, and suggest that some study of those constraints relevant to a particular problem should probably be made before seriously contemplating use of a sequential allocation scheme.

Remarks (i). It seems to be an interesting mathematical problem to explain the success exhibited in Table 1 for the approximations (14) and (15). Heuristic arguments indicate that these approximations should be valid to within $0(1)$ as $b \to \infty$, $\nu \to \infty$, and $b\nu^{-1} \to$ const. But proving this result or, what is more interesting, determining the constant implicit in the $0(1)$ may be rather difficult.

(ii) A challenging problem is to extend Proposition 1 and its consequences to other situations. An interesting discussion by Jennison, Johnstone, and Turnbull (1981) shows that the naive generalization to three populations is not valid.

REFERENCES

Bather, J. A. (1980). Randomized allocation of treatments in sequential trials, Adv. Appl. Probab. **12**, 174-182.

Bather, J. A. (1981). Randomized allocation of treatments in sequential experiments, Jour. Roy. Statist. Soc. B, **43**.

Blackwell, D. and Hodges, J. L. (1957). Design for the control of selection bias, Ann. Math. Statist. **28**, 449-460.

Flehinger, B., Louis, T. A., Robbins, H., and Singer, B. (1972). Reducing the number of inferior treatments in clinical trials. Proc. Nat. Acad. Sci. USA **69**, 2993-2994.

Hayre, L. S. (1979). Two population sequential tests with three hypotheses, Biometrika, **66**, 465-474.

Jennison, C. Johnstone, I. M., and Turnbull, B. W. (1981). Asymptotically optimal procedures for sequential adaptive selection of the best of several normal means, Proceedings of the Third Purdue Symposium on Statistical Decision Theory and Related Topics, S. S. Gupta and J. Berger, Eds.

Louis, T. A. (1975). Optimal allocation in sequential tests comparing the means of two Gaussian populations, Biometrika, **62**, 359-369.

Robbins, H. (1974). A sequential test for two binomial populations, Proc. Nat. Acad. Sci. USA, **71**, 4435-4436.

Robbins, H. and Siegmund, D. (1974). Sequential tests for involving two populations, J. Amer. Statist. Assoc. **69**, 132-139.

Siegmund, D. (1979). Corrected diffusion approximations in certain random walk problems, Adv. Appl. Probab. **11**, 701-719.

Research partially supported by Office of Naval Research Contract number N00014-77-C-0306.

NON-DETERMINISTIC MODELLING AND ITS APPLICATION IN ADAPTIVE OPTIMAL CONTROL

Ian H. Witten

Department of Computer Science, University of Calgary, Calgary, Canada T2N 1N4

Introduction

A problem which occurs in every branch of system theory is the conflict between data acquisition for system identification and system control towards some objective. The appropriate actions for identification are rarely those most appropriate for control, and yet in most real system studies control actions cannot be determined properly without some prior identification. The need to balance the costs of further exploratory activity against the costs of poor system knowledge creates a basic conflict that cannot be avoided. This conflict is succinctly expressed in the two-armed bandit problem, which has received a great deal of attention since it was formulated three decades ago.

This paper discusses how a two-armed bandit controller (in reality, an m-armed bandit controller -- the generalization is usually straightforward) can be employed as a basic instrument to solve more complex system-theoretic problems. We first consider how to build controllers for a class of discrete-time stochastic environments, which includes the two-armed bandit controller as a special case. The controller receives the environment's current state and a reward signal which indicates the desirability of that state. In response, it selects an appropriate control action, notes its effect; and the cycle repeats indefinitely. It is structured as an ensemble of m-armed bandit controllers interconnected into a network. (The design of these components is not considered here.) It is necessary to introduce probability estimators which provide interactions between the local optimization tasks accomplished by the m-armed bandit controllers. Under certain conditions, the controller's actions eventually become optimal for the particular control task with which it is faced, in the sense that they maximize the expected reward obtained in the future.

It is extremely interesting that global optimality of the control policy can be obtained from locally optimal m-armed bandit controllers. However, a key assumption is that the environment's current state is always available to the controller. In most realistic applications it will be necessary to estimate the state of the environment by observing its outputs and from them forming a model of it. Although there is a well-developed theory of observability for linear systems, the problem of modelling non-deterministic, non-linear systems from observations of their behaviour is still poorly understood. Even the formulation of the problem is non-trivial. Given a finite stretch of behaviour from a non-deterministic generator, complex models of it can be made which do not necessarily illuminate the structure of the generator. A formulation in terms of a trade-off between *model complexity* and *goodness-of-fit* to a particular behaviour sequence allows good models to be found by enumeration. Unfortunately this is extremely computation-intensive, and so weaker modelling methods which do not suffer exponential growth in computation are of interest. It is an open question how an optimal controller which assumes perfect state information will perform using an imperfect model found inductively from observation of behaviour.

The organization of the paper is as follows. We begin by formulating the control task in rather a general way. This task constitutes the "environment" of the controller. It is assumed that the controller initially knows nothing about the structure of the environment (except that it operates in discrete time), and is charged with a general goal of maximizing one of the environment's outputs - the reward output. After this initial formulation of the problem, the body of the paper is in two main sections.

In the first it is assumed that the environment provides a "state-output" which keeps the controller informed as to which state it is in. A controller structure is presented which identifies the state-transition characteristics of the environment and the areas of high reward, and gradually adopts a control policy which is optimal under the given reward-maximization criterion. Theorems are given (and proved elsewhere) which show that the controller cannot be in equilibrium unless it is operating under an optimal policy, and that when it is not in equilibrium any changes in its policy will tend to be improvements rather than degradations.

The second main section considers the problem of providing the "state-output" needed above from observation of auxiliary outputs, which do not provide state identities explicitly. The problem is formulated as one of non-deterministic structural modelling. Three distinct approaches to this which have emerged in the last decade are surveyed: enumeration and evaluation of candidate models (which produces excellent results at the expense of a huge amount of computation); reduction and evaluation; and limited-context methods (which give good results for a restricted class of environments).

The environment

The controller operates in a task environment which comprises a finite-state stochastic automaton. At each state the environment provides a reward value which is characteristic of that state, and is received by the controller. The controller's task is to maximize the total expected reward over a long period of time. It can exert control over the environment by manipulating control inputs to it; however, nothing is known in advance about the likely effects of these.

The environment is modelled as a stochastic automaton with n states 1, 2, ..., n. When in any particular state, it emits a reward of $g \in [0,1]$, accepts one of m control actions 1, 2, ..., m from the controller, and changes to a new state. It is assumed that every state is reachable, albeit indirectly, from every other. The environment operates in a discrete-time world and is synchronized with the controller.

The reward output is a random variable whose mean value indicates how well the controller is doing. We assume that this mean value characterizes the current state of the environment (Moore model). If necessary, it is easy to re-cast the formulation to take account of environments where the reward depends on the last transition between states (Mealy model), rather than the last state alone. The environment is characterized by the $n \times n \times m$ matrix p, where

$p_{ijk} = Pr$[environment goes from state i to state j under control action k];

and the reward expectation vector g, where

$g_i = E$[reward value when the environment is in state i].

None of this information is available to the controller; it must be inferred from observations of the environment. The dynamic behaviour of the environment is assumed to be captured by the state-transition model, and the random nature of the rewards simply represents noise in the rewarding process. Thus successive reward outputs are assumed to be statistically independent.

The controller's task is made much easier if it always has knowledge of the current state of the environment. In the first part of this paper we will assume that the environment makes available a state-output i, which indicates the state that it is in. These state-outputs can be viewed as *markers* which show the controller when the environment has reached a point which is equivalent to previous points in the interaction. Notice that this by no means makes the controller's problem trivial, for it must still model the dynamic behaviour of the environment and relate it to the rewards received, before it can issue good control actions. In the second half of the paper we will discuss how to dispense with the rather unrealistic assumption that state outputs are provided.

A *control policy* for the environment is an $n \times m$ stochastic matrix π, where

$\pi_{ik} = Pr$[when the environment is in state i, control action k is taken].

The controller's aim is to find a policy which maximizes the total expected reward over a long sequence of transitions. The choice of control action need not depend on anything other than the current state, since the environment obeys the Markov property. A *discount factor* $\gamma \in (0,1)$ is introduced in order to weight the immediate future more heavily than the distant future. If a sequence of rewards h_0, h_1, h_2, ..., h_r is obtained in the future, beginning with the present reward h_0, the *discounted reward* is defined to be

$$(1-\gamma) \sum_{s=0}^{r} \gamma^s \, h_s \; .$$

By manipulating γ, we can make the controller's goal short-term or long-term optimization.

Let $d_i(\pi,r)$ denote the expected discounted reward after r transitions, starting at state i of the environment, under policy π. A control policy π is said to be *optimal* if, for any other control policy π',

$$\lim_{r\to\infty} d_i(\pi,r) \geq \lim_{r\to\infty} d_i(\pi',r) \quad \text{for all } i \in 1, 2, \ldots, n.$$

Clearly, a policy can only be optimal if π_{ik} is either 0 or 1 for each i and k -- unless, by chance, there is a state i and two actions k and k' each leading to exactly the same expected discounted reward.

The controller

The controller to be presented is synthesized out of n "elementary controllers", which are called MABCs (m-armed bandit controllers), one for each state of the environment. To provide motivation for the introduction of MABCs, consider the special m-state, m-control-action environment with transition matrix

$$P_{ijk} = \delta_{jk} \quad (\delta \text{ is the Kronecker delta}).$$

Here, action k always takes the environment to state k, regardless of the previous state. The optimal control policy consists of exclusively selecting actions k^* for which

$$g_{k^*} = \max_k \{g_k\}.$$

This is the m-armed bandit problem, an obvious generalization of the familiar two-armed bandit problem which has been discussed extensively in the literature (Witten, 1976).

The control environment differs from a two-armed bandit in two ways. Firstly, the number of allowable control actions may exceed two. This leads to the m-armed bandit problem, which can be modelled in terms of m coins of different, unknown, biases; and an m-way choice at each trial. The extension of two-armed bandit strategies to the m-armed case normally presents no more than technical difficulties. Secondly, and more importantly, control environments like the general one introduced in the previous section have many states, and the best action to take usually depends on the state. One is naturally led to consider the environment as an ensemble of m-armed bandit problems. However, these constituent elements are not independent. We do not seek to optimize the *immediate* output, but instead have a performance criterion which usually depends on the results of interactions with many of the elements, namely, the discounted reward. And the control action which is best in the long term will not necessarily optimize the immediate output. One can visualize a situation with a valley between two peaks, where the controller must be prepared to leave one summit and traverse the valley if the other peak is high enough. The construction of such a controller is discussed in the next section.

As noted above, we will use m-armed bandit controllers as components of the adaptive controller to be developed. We will not discuss the m-armed bandit problem as such, but assume that a suitable solution to it is given. The real problem we face is how to connect such controllers together to make an optimal controller for a general Markov environment. This strategy, of assuming a suitable MABC to be given, will at least ensure that our controller performs well on the restricted type of environment described at the beginning of this section.

Control with a network of MABCs. The controller will comprise, for any state i of the environment:

(i) an MABC, denoted $MABC_i$, which will be called upon to select the controller's action when the environment's state-output is i, and which will be rewarded after this selection according to its success;

(ii) an estimate e_i of the expected discounted reward obtainable when the environment is in state i.

Suppose the environment is in state i, and $MABC_i$ selects control action k, which causes the environment to change its state to j and produce reward output g. Then e_i will be updated by

$$e_i \leftarrow (1-\beta)e_i + \beta g' \quad (\beta \in (0,1) \text{ is a constant}),$$

where g' is a weighted average of the environment's reward g and the controller's new estimate of future reward:

$$g' = (1-\gamma)g + \gamma e_j \quad (\gamma \in (0,1) \text{ is the discount factor}).$$

g' is also used as a "computed reward" to reinforce $MABC_i$ for its choice of control action k. When the controller has completed these updating operations, it will call upon $MABC_j$ to select the new control action, and the cycle will begin again.

Each of the n MABCs is characterized, at any one time, by the m-vector of probabilities with which it will choose the control actions. When these vectors are put together as rows of an $n \times m$ matrix, they form the instantaneous control policy π for the controller. Thus the state of the controller comprises its policy matrix π together with the vector e of current estimator values. We denote this state by $\langle \pi, e \rangle$.

Equilibrium states and mean states. Suppose the controller's policy is π. This induces a transition matrix τ on the environment:

$$\begin{aligned} \tau_{ij} &= Pr[\text{environment goes from state } i \text{ to state } j] \\ &= \sum p_{ijk}\, \pi_{ik}\,. \end{aligned}$$

Now we can compute the mean value $\hat{e}_i$ of the estimator e_i, given π.

$$\hat{e}_i = \underset{j}{E}[(1-\gamma)g_j + \gamma e_j],$$

where the expectation is taken over all states of the environment which can follow state i under policy π.

$$\begin{aligned} \hat{e}_i &= \sum \tau_{ij}[(1-\gamma)g_i + \gamma E[e_j]], \\ &= \sum \tau_{ij}[(1-\gamma)g_i + \gamma \hat{e}_j]. \end{aligned}$$

Therefore,

$$\hat{e}(\pi) = (1-\gamma)(1-\gamma\tau)^{-1}\, \tau g.$$

Note that $(1-\gamma\tau)^{-1}$ exists, since τ is a stochastic matrix and so all eigenvalues of $\gamma\tau$ must be strictly less than 1 in absolute value.

The above equation for e_i shows that it is exactly the $\lim_{r\to\infty} d_i(\pi,r)$ mentioned earlier, where $d_i(\pi,r)$ is the expected discounted reward obtained after r transitions, starting at state i, under policy π. Hence the limit exists, and the d_i's can be calculated from the control policy and the parameters of the environment, using the earlier equations.

The ith MABC is in an *equilibrium state* if its expected reward cannot be improved immediately by a change in its policy alone. This definition is motivated by the similar concept of an equilibrium state of a game (Nash, 1951). Indeed, it seems at first sight that some of the results concerning equilibria of games are immediately applicable to our controller, since the ensemble of MABCs can be considered as an n-person game with outcomes determined by the control environment. Unfortunately, this is not the case: a fundamental assumption in game theory is that the outcomes -- our e's -- are linear functions of the mixed strategies of the players, but the development above shows clearly that $\hat{e}$ is not a linear function of π.

The state of the controller, $\langle\pi,e\rangle$, can alter in two ways. Firstly, the control policy π can change. This corresponds to a change in the state of one or more of the constituent MABCs, and will presumably not occur unless the relevant MABC is not in an equilibrium state. Secondly, the estimates e_i can change. However, if π remains constant, the expected values of these estimates will approach $\hat{e}(\pi)$ exponentially at a speed which depends on the constant β used for updating the estimates. Thus we call states $\langle\pi,e\rangle$ with

$$e = \hat{e}(\pi)$$

the *mean states* of the controller. If the system is started in a mean state, it will fluctuate about that state in a random manner, but the expected values of the estimators will not change.

Mean states are defined in quite a different way to equilibrium states. The former refers to the controller as a whole -- the ensemble of MABCs -- and simply requires the estimators to have settled down under the current control policy. The latter refers to an individual MABC, and requires that it, in isolation, cannot improve upon its expected reward. One could easily imagine that a controller could have all of its constituent MABCs in equilibrium but still not be operating under an optimal policy, because although each state is at a local optimum, this does not necessarily seem to imply that the ensemble has reached a global optimum. However, the construction of the estimators e_i is designed to spread the effect of individual decisions by MABCs around the whole network. The fact that they do so is testified by the following result (Witten, 1977).

THEOREM 1. Suppose the controller is in a mean state $\langle\pi,\hat{e}(\pi)\rangle$, and each of its constituent MABCs is in equilibrium. Then if $\langle\pi',\hat{e}(\pi')\rangle$ is any other mean state,

$\hat{e}_i(\pi') \le \hat{e}_i(\pi)$ for all i.

In other words, the control policy π is optimal in the sense defined above.

Goal-directedness. We have seen that the local optimality of the MABCs implies global optimality of the controller as a whole, assuming that the estimators have settled down in the sense that the controller is in a mean state. This result by itself does not guarantee that a suitable control policy will be found, however; only that if one is stumbled upon by accident it will not be left. (Actually, since the estimators will fluctuate statistically, it may well be left.) A stronger result is needed to ensure that the system searches actively for a good control policy, rather than wandering blindly until it finds one.

The following result can be shown about the changes in policy as each of the MABCs adapts to its local environment, under the assumption that this adaptation takes place much more slowly than that of the estimates e_i (Witten, 1977).

THEOREM 2. Suppose the controller is in a mean state $<\pi,\hat{e}(\pi)>$. If one or more of the MABCs changes its policy in the direction of increasing immediate computed reward, giving a new policy matrix π', then

$\hat{e}_i(\pi') \ge \hat{e}_i(\pi)$ for all i.

Since optimality corresponds to maximization of the $\hat{e}_i(\pi)$'s, this result indicates that as long as the MABC's behave sensibly, only changing their policy in the direction of increasing expected reward, then the controller as a whole will converge on the optimal policy. By Theorem 1, once that situation is reached it will not be left (except as a consequence of statistical fluctuation in the estimators).

Effects of statistical variance. The above analysis assumes that the controller is in a mean state, and neglects fluctuations about mean states. An examination of the effects of variance in the controller show that its parameters can be chosen to make it unlikely that the inevitable random perturbations of the estimates will cause serious corruption of an optimal control policy (Witten, 1977). Of course, if such corruption does occur it will not be permanent, for the controller always works back towards optimal policies.

Structural identification to provide state-information

In the last section we assumed that the environment provides a state output which indicates its state at every stage. This information is absolutely necessary for the scheme to work, for the MABCs are introduced on a state-by-state basis, each being associated with one state of the environment. In actual applications, however, it is unlikely that such information will be provided. Although the environment may generate an auxiliary output as well as the reward signal, this output will probably not identify its state uniquely. Rather, it will be produced by an output function

λ: state set $\to$ output set

which may be deterministic or stochastic. Then the controller will select a control action c, and observe the environment's response o and the reward output g. From this, if the control scheme presented above is to work, it must deduce a state s in order to identify the MABC to be called into play to select the next control action. Hence given a sequence

$$c_1 \begin{bmatrix} o_1 \\ g_1 \end{bmatrix} c_2 \begin{bmatrix} o_2 \\ g_2 \end{bmatrix} c_3 \begin{bmatrix} o_3 \\ g_3 \end{bmatrix} \dots ,$$

a corresponding sequence of states

$$s_1 \quad s_2 \quad s_3 \quad \dots ,$$

must be identified. We will call the first sequence a *behaviour sequence*

$$b_1 \; b_2 \; b_3 \; b_4 \; b_5 \; b_6 \; \dots ,$$

with elements b_i drawn alternately from the set of control actions and the cross product of the output and reward sets. Then the state identification problem becomes one of behaviour-to-structure identification: given the behaviour sequence, it is necessary to determine a state model which generates it.

It is quite possible that the environment generates no output at all, other than the reward. In this case the behaviour sequence will contain alternate control actions and rewards. On the other hand, it may be that the outputs are quite good indications of the state of the model, while the rewards have a high variance. In this case it may well be desirable to omit the reward component from the behaviour sequence, so that it comprises alternate control actions and environment outputs. This is because increasing the cardinality of the set from which behaviour symbols are drawn has a devastating effect on the computational complexity of some behaviour-to-structure identification techniques.

Formulation of the structural modelling problem. It is simple to form a causal model of a behaviour sequence if the model is permitted to be as large as the deterministic automaton that generated the sequence. However, if there is a limit on the number of states allowed in the model, or if the sequence is noisy or generated non-deterministically, the problem of optimal causal modelling becomes very difficult indeed. Under these circumstances the model must be non-deterministic.

Three distinct methods for non-deterministic structural modelling have been introduced in the last decade and will be reviewed here: enumeration and evaluation, reduction and evaluation; and limited-context. To summarize their properties, the first method guarantees to produce an optimal model of any pre-specified complexity, for a given criterion of "optimality". The price paid is extraordinarily long computation time, which grows super-exponentially with the model complexity. Reduction methods are usually based on heuristic criteria, which makes them difficult to evaluate properly. However, we will present below an economical method of reduction which can be based on any given optimality criterion. It does not necessarily produce optimal models, but it does give ones which are sometimes much better than those found by other methods, in regions where the optimal model is unknown because enumeration cannot reasonably be attempted. Finally, limited-context methods are extremely fast, but can only deal with (non-deterministic) input strings that are "k-testable", and this excludes behaviours which involve counting (modulo an integer) symbols that may be separated by arbitrary strings of other characters.

A methodology for the structural modelling problem has been developed by Gaines (1976b, 1977). He defines the subset of *admissible models*, given a measure of poorness-of-fit of a model to the behaviour sequence and a criterion of model simplicity. This admissible subset contains only models which cannot be improved in simplicity without increasing poorness-of-fit, and cannot be improved in goodness-of-fit without increasing complexity. The admissible subset is found by enumerating all models in increasing order of complexity, and evaluating each one with respect to the behaviour sequence using the poorness-of-fit measure. If a *probabilistic* model is desired such an enumeration may appear to be impossible, for the space of candidate models is continuous rather than discrete. A key result of Gaines is that this is not so, for an optimal assignment of probabilities to the branches of a non-deterministic model may be made by running the behaviour sequence through the model and making frequency counts.

We will adopt here the number of states in a model as the measure of its complexity. It seems plausible that the convergence speed of the adaptive control algorithm developed earlier will be directly related to the size of its environment model. For a given duration of interaction, the states of a small model will, in general, be visited more often than those of a large one. Since the probability estimators and MABCs are attached to the states, they will be updated more frequently and hence converge faster.

Enumerative modelling. Gaines's ATOM system enumerates models in order of increasing complexity and evaluates their goodness-of-fit, selecting for the admissible subset the best fit at each level of complexity. Enumeration is performed in a data-directed fashion by starting at the beginning of the behaviour sequence and building a model which fits it (in a causal, though non-deterministic, sense). In order to economize on search time, it is preferable to examine only those models which contain no unreachable or equivalent states and fit the behaviour sequence.

Biermann (1972) used the same method to infer Turing machines from traces of their execution. He, however, was concerned only with *deterministic* machines, with an input which represented the current symbol read from the tape. The assumption of determinism reduces the search space enormously. In the present work, however, we assume that the model operates non-deterministically. Any procedure which produces a deterministic model of a non-deterministically generated sequence will fail in the spectacular manner described by Gaines (1976a). This point is discussed further in Witten (1981).

The ATOM procedure enumerates non-deterministic models using a recursive search (Gaines, 1976). It evaluates each model as it is enumerated. For Biermann, evaluation is unnecessary because the enumeration only produces a "correct" model of the sequence. ATOM runs the sequence through the model and notes frequency counts, using these to compute the overall goodness-of-fit through an entropy (or some other string-approximation) measure.

The number of models examined by ATOM grows very quickly with the size of model. For example, there are n^{nq} Mealy models with n states and q behaviour symbols. Many of these will be ruled out by a particular behaviour sequence because they are unreduced, unreachable, or unobservable; however, the super-exponential behaviour as n increases, for fixed q, remains. Some advantage can be gained by searching the model space in a different order. Witten (1980) has investigated a special class of "transitive" Moore models which require substantially less computation (bounded by $(n+1)^n$) in most cases, without sacrificing the goodness-of-fit of the resulting model. It would be interesting to know whether other gains are possible while retaining the full power of this modelling technique.

Reduction and evaluation methods. One way of avoiding the exponential search required by ATOM is to employ a method of reduction, starting with a large model and successively merging states until the desired model size is reached. Evans (1971) applied this to the inference of picture grammars from examples. Beginning with an initial grammar containing production rules for each of the example objects, he collapsed this by merging states according to some heuristic criteria, amongst which were

- eliminate multiple occurrences of a rule
- identify pairs of non-terminals to produce multiple occurrences of a rule
- add productions of the form <non-terminal> → <string of terminals> to give multiple occurrences of a rule.

It is difficult to assess these heuristic rules according to the quantitative evaluation criterion used by ATOM. However, a modelling procedure based on reduction can easily be defined by taking a large model of a behaviour sequence and repeatedly coalescing pairs of states. The initial model could be a direct representation of the sequence itself (that is, an N-state model of an N-element sequence) if it is short. Other methods of quickly generating large, approximate models have been developed (Witten, 1979) and are discussed below. The most straightforward way to determine which states should be merged is to evaluate separately the models which result from merging each pair of states and select that giving the best evaluation. Suppose we begin with an N-state initial mode. Then the number of models which need to be evaluated to produce the (N-1)-state model is ${}^{N}C_{2}$. To create a spectrum of models with N-1, N-2, ..., 2, 1 states by deriving each from its predecessor takes $N(N^2-1)/6$ evaluation operations -- a polynomial rather than exponential complexity.

Instead of successively merging states in a pairwise fashion, one could generate an n-state model directly from an N-state one ($n<N$), without examining any model of intermediate size, by identifying the best set of N-n states to coalesce. To do this, ${}^{N}C_{n}$ models would need to be considered. In fact, to generate the full set of 1-state, 2-state, ..., (N-1)-state models from an N-state one by this method would require evaluation of 2^{N-1} models -- a return to exponential complexity. However, for short behaviour sequences this could lead to an improvement on the ATOM method. If the large, starting model represented the complete behaviour sequence, admissible models would certainly be generated. If one were to examine a sequence of 100 items, the entire set of admissible models could be found by reduction if $2^{100}-1$ were examined. The complexity of the ATOM method grows as n^{nq} for n-state models with q different symbols: even for q=2 this approaches 2^{100} when n is 13. Hence for non-deterministic models with 14 or more states, this optimal reduction procedure could require less search than ATOM.

Limited-context methods. Limited-context methods of forming a structural model of a behaviour sequence assume that the structure can be characterized by the set of consecutive k-tuples of symbols which occur in the sequence, for some limited "context" length k. Different techniques have been developed by Biermann and Feldman (1972), Klir and his colleagues (Klir 1975, 1976; Uyttenhove 1980), Andreae (1977) and Witten (1979).

Biermann and Feldman tackled the problem of approximately modelling an input-output sequence with an automaton which may have fewer states than the generator. While their technique is an excellent one for the purpose for which it was designed, namely providing a limited-complexity model of a deterministically-generated sequence, it fails to cope with true non-determinism in the manner described by Gaines (1976a). Faced with non-deterministic input, the models become more and more complex as k increases.

Several limited-context modelling methods effectively store information about all k-tuples which appear in the sequence, for some fixed value of k. Andreae's (1977) PUSS modeller uses a hash table for rapid access. All k-tuples are stored in the table as they occur. When a prediction is required, the k-tuples are retrieved whose first k-1 elements match the last-seen k-1 elements of the behaviour sequence. The k'th elements of these tuples are the various predictions. A disadvantage of this method is that large amounts of store can be occupied with non-deterministic transitions. In the extreme case, when modelling a random sequence whose elements are drawn from an alphabet of q symbols, q^k events will be stored in the hash table. Thus large areas of store are used to hold events which will be of little use for prediction anyway -- simply because they are non-deterministic.

Witten (1979) has investigated how k-tuples can be recorded economically by massaging them into the form of an automaton model. First the "ingenuous" model of k-tuples is constructed, with one state for each tuple and transitions which reflect the succession of tuples in the behaviour sequence. Only the last element of a tuple is recorded as the output of its state. The model is then subjected to a reduction process which coalesces states in a way that does not destroy information about the k-tuples which occurred in the behaviour. For example, if the random sequence mentioned above is long enough, the ingenuous model will have q^k states, but this will become a mere q states (for a Moore model) after reduction. Thus although an increased value of k normally provides a larger, more accurate, model; it does not necessarily do so if structure does not exist in the behaviour sequence -- as in our example. Note the contrast with Biermann and Feldman's method which will generate ever larger models as k grows, if the input is random.

An important feature of this method, which we call *length-k* modelling, is that reduction can be done *incrementally*, each new element of the behaviour sequence being incorporated into an already reduced model to form an updated version of it. This gives an advantage over other methods of non-deterministic modelling which require the complete input sequence to be stored, and re-model it from scratch when new information shows the current model to be inadequate.

The sequence of k-tuples which occur in a behaviour fully characterizes that behaviour. However, if the sequential information is discarded and just the *set* of k-tuples is stored, certain kinds of finite-state events cannot be modelled. McNaughton and Papert (1971) define *k-testable events* as ones which are completely characterized by the k-tuples which they can generate. They show that such events are *non-counting* in that the operation of counting (modulo some integer) cannot be performed if the elements to be counted can be separated by arbitrarily long sequences of other events. For example, the regular event $(A^*BA^*BA^*C)^*$ is a counting event, because two B's, which may be separated by arbitrary strings of A's, need to be counted before a C can be predicted. However, $(ABAABAAAC)^*$ is *non*-counting, because the separating strings are of fixed length.

Deterministic counting events can be identified by Biermann and Feldman's technique. However, if the event is non-deterministic, then structural identification is an extremely difficult problem. As far as I am aware, only exhaustive searching techniques like Gaines's ATOM and the optimal reduction method (both described above) are sufficiently powerful to cope with counting events.

Conclusions

This paper has assembled a toolkit of techniques for attacking the problem of adaptive identification and control of highly non-linear systems. Each of the main components has been tested separately, but they have not yet been brought together into a single system.

The adaptive controller developed in the first part of the paper has been shown to achieve an optimal control policy when the variance of the estimators is neglected and the estimators are assumed to converge rapidly compared to the MABCs. However, in practice the effects of nonzero variance and slow convergence may cause the control policy to deteriorate. The variance of the estimators can be kept low if

γ is small (short-term optimization)

β is small (slow estimating).

Bearing in mind the requirement that the estimators converge rapidly compared with the MABC's, the second condition requires that for reliable operation the time-constant of the MABC's must be extremely large. Thus the controller will take a long time to adapt to a new environment.

The performance of a controller constructed along the lines of the one described here has been studied experimentally (Witten and Corbin, 1973). Near-optimal control of a noisy third-order analogue plant was achieved consistently, but, as predicted here, the time taken by the controller to discover near-optimal policy was rather long. (In fact, it was typically 10,000 sampling periods, corresponding to about 30 minutes of real time.) However, using a general adaptive controller such as the one described here naturally incurs a penalty in convergence time. In general, control performance for particular tasks can (and usually should) be increased by building special-purpose constraints into the controller.

The non-deterministic identification techniques surveyed in the second part have been tested on a variety of sequences (see, for example, Gaines, 1976b; Witten, 1979). The enumeration and evaluation method, of course, guarantees good results. However, it is expensive. Notice that the problem here lies in a completely different direction from that of the controller itself: the controller requires a good deal of *plant time* to formulate a good policy for a stochastic environment, whereas enumerative identification needs large amounts of *computation time* to discover good structural models for a behaviour sequence. From our limited experience, it seems that only fairly short behaviour sequences are needed to give good approximations to the environment structure.

One outstanding problem with enumerative identification is that it cannot be done incrementally. If a new behaviour symbol is observed, re-modelling from scratch is necessary. Length-k modelling overcomes this problem, but is not guaranteed to produce good models for certain types of behaviour (namely, "counting" sequences). Furthermore, a length-k modeller needs lengthy samples of stochastic behaviours to construct faithful models, and the sample size needed grows exponentially with k, the desired complexity of the model.

While the techniques developed here are perfectly general, practical considerations such as those discussed above indicate that careful design will be needed if they are to be applied to real control problems.

Acknowledgements

It is a pleasure to acknowledge the influence of Brian Gaines and John Andreae, who have been a constant source of stimulation and encouragement to this work over many years. I am also most grateful to John Cleary for improving the presentation through many suggestions. This work is supported by the Natural Sciences and Engineering Research Council of Canada.

References

Andreae, J.H. (1977) *Thinking with the teachable machine*. Academic Press, London.

Biermann, A.W. and Feldman, J.A. (1972) "On the synthesis of finite-state machines from samples of their behaviour" *IEEE Trans Computers, C-21* (6) 592-597, June.

Biermann, A.W. (1972) "On the inference of Turing machines from sample computations" *Artificial Intelligence*, *3*, 181-198.

Evans, T.G. (1971) "Grammatical inference techniques in pattern analysis" in *Software Engineering*, edited by J. Tou, pp. 183-202. Academic Press, New York.

Gaines, B.R. (1976b) "Behaviour/structure transformations under uncertainty" *Int J Man-Machine Studies*, *8*, 337-365.

Gaines, B.R. (1976a) "On the complexity of causal models" *IEEE Trans Systems, Man and Cybernetics, SMC-6*, 56-59.

Gaines, B.R. (1977) "System identification, approximation and complexity" *Int J General Systems*, *3*, 145-174.

Klir, G.J. (1975) "On the representation of activity arrays" *Int J General Systems*, *2*(3) 149-168.

Klir, G.J. (1976) "Identification of generative structure in empirical data" *Int J General Systems*, *3* (2) 89-104.

McNaughton, R. and Papert, S. (1971) *Counter-free automata*. MIT Press.

Nash, J.F. (1951) "Non-cooperative games" *Annals of Mathematics*, *54*, 286-295.

Uyttenhove, H.J.J. (1980) "On the two-dimensionality in the behavioral system identification problem, Part 1" *Int J Man-Machine Studies*, *12*, 325-340.

Witten, I.H. and Corbin, M.J. (1973) "Human operators and automatic adaptive controllers: a comparative study on a particular control task" *Int J Man-Machine Studies*, *5*, 75-104, January.

Witten, I.H. (1976) "The apparent conflict between identification and control: a survey of the two-armed bandit problem" *J Franklin Institute*, *301* (1-2) 161-189, January-February.

Witten, I.H. (1977) "An adaptive optimal controller for discrete-time Markov environments" *Information and Control*, *34*, 286-295, August.

Witten, I.H. (1979) "Approximate, non-deterministic modelling of behaviour sequences" *Int J General Systems*, *5*, 1-12, January.

Witten, I.H. (1980) "Probabilistic behaviour/structure transformations using transitive Moore models" *Int J General Systems*, *6* (3) 129-137.

Witten, I.H. (1981) "Some recent results in non-deterministic modelling of behaviour sequences" *Proc Society for General Systems Research Annual Conference*, 265-274, Toronto, Ontario, January.